THE FINISHING TOUCH

A History of the
Texas Cattle Feeders Association
and Cattle Feeding in the Southwest

by Charles E. Ball

Texas Cattle Feeders Association
Amarillo, Texas

Library of Congress Catalog Card Number
92-61319

ISBN 0-9627761-2-2

Designed by
Gayle Allston
McCormick Advertising Agency
Amarillo, Texas

Printed by
Trafton Printing, Inc.
Amarillo, Texas

Published by
Texas Cattle Feeders Association
5501 West I-40
Amarillo, Texas 79106
(806) 358-3681

Contents

About The Author

Charles E. Ball is uniquely qualified to write a history of TCFA and cattle feeding in the Southwest.

As an award-winning editor of ***Farm Journal*** during the 1950s and 60s, he wrote articles about and promoted cattle feeding in the Southwest. He appeared on the program of the first TCFA convention in 1967 and has attended every convention since.

As Executive Vice President of TCFA, 1972-88, he fought the battles and helped celebrate the victories of cattle feeders . . . during the worst of times and the best of times.

Ball is a native of Lamar County, Texas, where he won many 4-H Club honors. He graduated from Texas A&M College (1947) and Iowa State College (1948), with degrees in agricultural engineering after a three-year interruption during World War II for the army.

His earlier writing experience included: editor of the ***Texas A & M Agriculturist*** (1947); managing editor of ***Farm and Ranch - Southern Agriculturist*** (1948-52); and Southwest editor of ***Farm Journal*** (1952-72). In addition, he has written for ***The Saturday Evening Post*** and ***Reader's Digest.*** Also, Ball was author of a "best seller" horse book, ***Saddle Up!***

ACKNOWLEDGMENTS

In reading this book, you're sure to remark – scores of times – "I didn't know that!" I said that many, many times while researching, digging and interviewing.

In fact, I've been researching this book – literally making and keeping notes – for more than 20 years, ever since I joined the staff of TCFA in 1972. And for 20 years before that, as a ***Farm Journal*** editor, I witnessed and marveled and reported the amazing developments in the cattle business, including cattle feeding.

But as I concentrated on the history of cattle feeding in the past 100 years in the Southwest, and the past 300 years in the United States, it became amusingly apparent: "Some things never change."

Also, I learned that little had been written on the history of cattle feeding. The bookshelves are loaded with books about cattle trail drives, famous ranches and cattle breeds. But none about "the other half" of the industry – cattle feeding.

As thorough as I have attempted to be, there remains much to be told. My only regrets are that time and space did not permit more recording of significant events and important contributions by notable people. Nevertheless, this book represents a conscientious effort to record the evolution, the trials and the glamour of cattle feeding; and it is the best information available at this point in time.

As always with a project of this scope, numerous people made valuable contributions in various ways. I am especially grateful to the following: Brenda Higley for loyal counsel and encouragement and help in many ways; Steve Cornett for help in research and writing; Kathy Cornett for editing assistance; Gayle Allston for creative design and layout; Ralph Leone and Jim Peck for expert photo processing; Carol Williams at the Texas and Southwestern Cattle Raisers Foundation, Clara Kuehn at the Panhandle Plains Museum and David Murrah at the Southwest Collection for their assistance in research; Brenda Reeder and Carla Walker for typing the manuscripts; my wife, Dawn, for her patience and encouragement; and, finally, Richard McDonald and the Texas Cattle Feeders Association for making the whole project financially feasible.

Some things never change. The value of friends – caring, loyal, helpful friends – is one of them.

Harper's Weekly

Cattle feeding started in New England over 200 years ago, as depicted by artist, then extended into the Midwest.

Historic Sketches of the Cattle Trade

"Premium Texas cattle," were fed in Illinois by John B. Hunter, among the first to corn-feed large lots of Texas cattle. These steers, shown at the 1873 Kansas City Exposition, weighed nearly 2,000 pounds and were pronounced "the fattest cattle ever seen."

Chapter 1

FEEDYARDS AND FOREFATHERS

Cheap grass, grain, brewers mash, turnips, beets and apples made long-feeding feasible in the 18th and 19th centuries.

During the last 25 to 35 years, the southern Great Plains observed the development of a booming new industry – immense in scale, impressive to the observer and important to the economy.

Commercial cattle feeding in Texas, Oklahoma and New Mexico grew from a negligible enterprise into a $14 billion industry!

The primary rationale for feeding cattle has been twofold: (a) to increase weight and (b) to improve quality grade. But early on, there was another reason – to utilize a surplus of feedstuffs.

And so it has been through the 300-year history of cattle feeding in the United States.

The Frontier of Cattle Feeding

The first cattle in the U.S. came with the Spanish explorers in the 1490s and were an important part of Spanish settlement in the Southwest. But cattle feeding is more a product of the European colonists, who sank their roots into the East Coast starting in 1620.

As trappers and explorers learned the country and began to move westward, they were followed by the earliest ranchers, eager to take advantage of the free grazing. The "western frontier" to those early settlers often was only a hundred or few hundred miles from their roots on the East Coast. But a pattern was established early that would eventually move across the continent.

First would appear the explorers; then the ranchers to introduce and multiply cattle numbers; then farmers to break the sod and produce crops; then cattle feeders to combine the cattle and the crops into a sometimes-profitable venture.

So whether the moving frontier was in western New York or Ohio or Illinois or Iowa or Nebraska or Texas, the early pattern of development was nearly always the same: first, explorers; second, ranchers; third, farmers; and fourth, cattle feeders.

The early cattle feeders would take advantage of the cheap supplies of cattle to their west, the abundant supply of feed locally and move their finished cattle to urban markets in the East.

For instance, at the time of the American Revolution (1774-1783), southeastern Pennsylvania was an important feeding area. Cattle purchased from cow herds in the Carolina Piedmont and the Shenandoah Valley were fattened on Pennsylvania corn, then driven to market in cities such as Charleston, Baltimore or Philadelphia.

Around 1800, cattlemen in the Hartford, Connecticut area were "finishing" several thousand head of steers a year on brewers mash, root crops and apples, the steers being tethered to the trough or confined in small stalls. Hartford Distillery, one of 21 distilleries in the area, fed as many as 1,000 head a year in this manner.

Fat Cattle Driven to Market

By the early 1800s, farmers in the Bluegrass Region of Ohio and Kentucky were feeding cattle. The first fat cattle from Ohio were driven to eastern markets about 1805. A NewYork City paper reported on an 1817 drive by a Mr. Drenning from Chillicothe, Ohio, a distance of nearly 1,000 miles. The paper said Drenning had left Ohio with about 200 head of cattle, sold many of them on the way and arrived in June. Local butchers were offering $12 per head, according to the paper, but the owner was asking $12.50. By 1840, Ohio had 1.2 million head of "neat" cattle (for slaughter), ranking it second in cattle numbers only to New York.

Professional drovers could buy cattle for 2 cents to 5 cents per pound, and it cost $12 to $15 per head to drive them from Kentucky or Ohio to Philadelphia or New York.

That arithmetic might not add up to an outsider – selling finished steers for less than total costs – but some entrepreneurs, then as now, always seemed willing to assume the risk, in hopes of respectable profits.

Some drovers fared better. That same year, 1817, drover Felix Renick reportedly got $133 per head for cattle in Philadelphia, netting $77 per head.

THREE CENTURIES OF BEEF EXPORTS

Today's cattle feeders are gaining a new appreciation for beef exports, which in 1990 accounted for approximately 5 percent of the tonnage and 10 percent of the value of U.S. production. A Cattle-Fax study showed that exports that year increased fed cattle prices $4.76 per cwt.

The U.S. Meat Export Federation was formed in 1976; and since that time, beef exports have increased 15 fold (in value), to $1.8 billion in 1991!

Those figures might or might not impress cattlemen of two and three centuries back. For imports were important in that era, too. Following are some sketchy entries from journals back then:

In 1652, farmers in the Connecticut Valley were stall-feeding cattle, which they slaughtered and made ready for shipment to the West Indies.

In 1660, Maverick wrote, "to consider how many thousand neat beasts (cattle for beef) and hogs are yearly killed and have been for many years past and sent abroad to supply Newfoundland, Barbados and Jamaica."

In 1743, Sir William Gooch, Colonial Governor of Virginia, wrote to the British Board of Trade and the King of England that during the year 20,000 barrels of pickled beef had been exported to Portugal and the Isle of Madeira. Another 20,000 barrels had been exported to the British West Indies, "plus 90,000 pounds of fresh beef."

In 1789, Brissot wrote that Boston sent 44 vessels to the Pacific northwest coast and the Orient. Examples he gave of Massachusetts' agricultural exports for 1790 included: horns and horn tips, 71,281; horned cattle, 652; salt beef, 30,499 barrels; fresh beef, 92,269 pounds.

In 1804, the ship "Juno" left Bristol, Rhode Island, for Asia with, among other commodities, 3,800 pounds of salt pork and 8,420 pounds of salt beef.

So given the population, the technology and the livestock numbers of that era, beef and cattle exports perhaps were as important to cattlemen of the Colonies as they are today. In fact, between 1868 and 1885, exporters found profitable the shipping of over 800,000 cattle valued at more than $80 million, an amount representing about 6 percent of the total value of beef products in the U.S. during that period.

Brighton Market in Boston was reputed to be the first livestock auction in the U.S., beginning about 1800. After the Western Railroad arrived in 1842, as many as 8,000 cattle might pile up for the weekly market.

Farmers typically bought three-year-old to five-year-old steers, then fed them not in feedpens but in fields that needed manuring. A wagon was used to haul shocked corn to the feeding field, and the farmer attempted to move the feeding evenly over the field to assure even distribution of the manure.

The older steers were preferred for two reasons. First, it was believed that cattle raising logically fell into two phases – the first phase for growing and the second phase for fattening – a belief that prevailed into this century. The second reason for the older steers was that they could better survive the long drives, which sometimes stretched up to 1,000 miles and lasted four to six months.

Before railroads and refrigeration, livestock and corn whiskey were about the only ways many farmers could market their grain. The livestock could walk to market, and whiskey was worth the transportation expense.

Each eastern city had its own market and beef packing industry for fresh meat, but most of the beef and pork was pickled (salted) and shipped in barrels. Much was exported, even during colonial times.

First Meat Packers

The 19th Century would see many changes in cattle feeding. Prior to the Civil War, Illinois was fairly entrenched as a feeding area. Chicago's first meat packing business was started by merchant George W. Dole in 1832 and by 1849 the ***Prairie Farmer*** declared Chicago "the most important packing center in the country."

Early, the Chicago packers used ships on the Great Lakes to transport beef and pork to New York markets. After the arrival of railroads in the 1850s, the rails were used. Following the Civil War, the packing industry flourished in Chicago – 2.2 million head were slaughtered by 1890.

The packers did most of their work in the winter. But by the 1860s, they were experimenting with ice to refrigerate beef on long hauls. Mechanical refrigeration was introduced in the 1890s.

Mechanization was beginning to have an impact, too. The corn sheller (invented about 1840) and hammer mills were starting to change the way cattle were fed – from

Harper's Weekly

Harper's Weekly

From 1868 to 1872, four packing plants were built in Kansas City, the largest and most modern being Plankinton & Armour, which slaughtered up to 1,000 cattle and 3,000 hogs a day (top). Cattle were shot outside and pulled inside with a windlass (bottom).

Harper's Weekly

The National Provisioner

Arrival of railroads (above) during latter part of 19th Century ultimately ended the long cattle trail drives.

Chicago Union Stockyards (left), depicted by an artist soon after it opened in 1865. It ceased operations in 1971.

Historic Sketches of the Cattle Trade

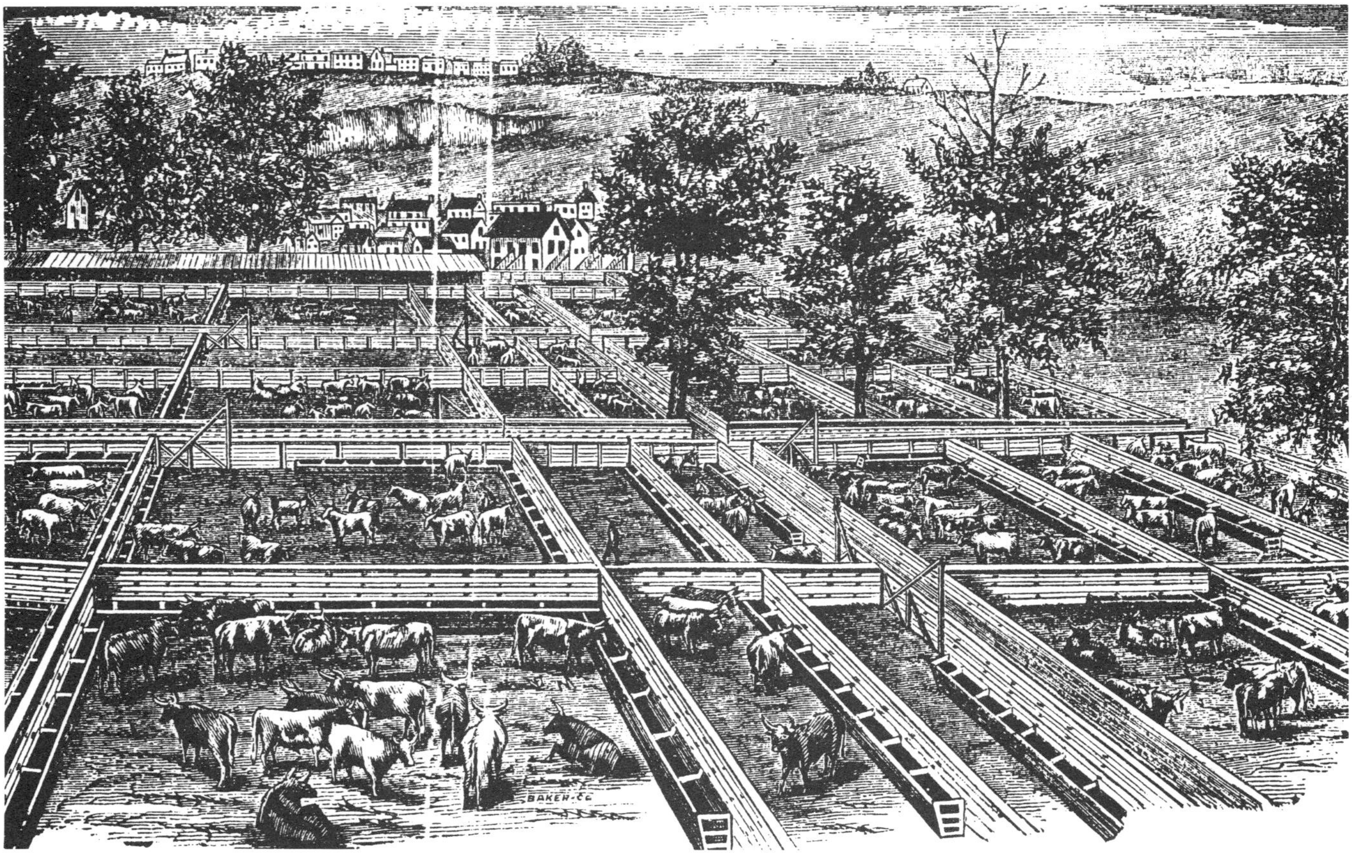

Kansas City Stock Yards, which grew into a successful market in 1872, became a popular spot for western cattlemen to meet and trade with eastern buyers who shipped by railroad. The Kansas City Stock Yards closed in 1991.

shocked corn in the field to shelled and ground corn in troughs – thus improving feed efficiency.

Following the Civil War, the number of cattle in the U.S. was growing rapidly – doubling from 1867 to 1890. This caused some significant market crashes, even by today's standards. National average cattle prices dropped from $25.26 per head in 1884 to $16.95 per head in 1890. Platform scales were uncommon, so it simplified trading to sell by the head.

By-Products for Feed

Around the turn of this century, sugar mills, whiskey distilleries and cottonseed oil mills were feeding significant numbers of cattle – using their by-products for feed.

For instance, Jimmy Moore of Harrodsburn, Kentucky tells about early-day feeding in Kentucky: "Jake Graves of Lexington started feeding brewers mash to oxen as early as 1888. After the oxen were worn out as draft animals, Mr. Graves would buy them, chain them to the trough and they'd fatten fast.

"With 40 to 50 distilleries in eastern Kentucky, cattle feeding boomed," continues Moore, "until the state was voted dry in 1916. . . . Then it picked up again in 1932, when Prohibition ended."

With feed so plentiful, there was little pressure during the 1800s to hurry cattle. In fact, if the price didn't suit them, ranchers often held their entire herds off the market for a year or more – their way of leveling out "cattle cycles."

Such long feeding periods were possible because grass and corn were so cheap in the farming areas, as were root crops. Turnips, rutabagas, beets, sweet potato vines, cowpeas, peavines and other locally popular feedstuffs were also used in some circumstances through the years.

Corn was typically shocked and fed as a whole plant – on the ground in fields or in large pens. Two to three hogs were run in the same pens for each beef animal being fed, and thus pork and beef production grew side by side.

By the start of the Civil War in 1861, Illinois was firmly entrenched as a feeding area, and drovers were reaching out far to the West for feeder cattle – as far as Texas.

Cattle feeding remained important in Illinois for many years; but it was Iowa that in 1880 would become the biggest cattle feeding state, a position it would hold for almost a century.

Arrival of Railroads

The first railway hit Iowa in 1855, opening access to Chicago's new packing industry and the older markets in the East. By the start of the Civil War, farmers in eastern Iowa had begun feeding cattle, shipping the fed animals by rail.

In 1863, Iowa railways shipped less than 70,000 head of cattle east. By 1888 they handled more than 600,000. Receipts at the terminal market in Chicago, just 48,000 in 1857, were 2.6 million by 1896.

Meanwhile, the range industry moved farther west. Nebraska, Texas, Colorado, western Kansas were opened. U.S. beef cattle numbers shot from 12 million in 1867 to 50 million by 1890.

Texas ranked as the 17th most important cow state before the Civil War. But it moved to first place shortly after the War – increasing from 5.5 million head in 1867 to 9.5 million in 1890 – a position it would hold throughout the 20th Century.

THE DIFFERENCE BETWEEN PROSPERITY AND POVERTY

[Excerpted from ***Texas Live Stock Journal***, November 26, 1887]

Few people imagine how great a difference one-half a cent per pound in the price of beef makes to the cattle interests of the state of Texas. It is not the simple difference between profit and loss, but the difference between prosperity and bankruptcy.

Suppose that one man having 100 steers, 950 pounds average, and he sells as such cattle are selling now, the price being 2 cents, the net return is $19, the total return is $1,900. At 2 1/2 cents the net return is $2,350. The difference is $450. To the man who has 1,000 steers the difference is $4,500; and to the man or company having 10,000 steers the difference is $45,000.

The state of Texas manages to get rid of about one million alleged beef cattle each year. The cattle consist of fat yearlings of 650 and 700 pounds, to some very fine corn-fed cattle weighing 1300 to 1400 pounds. The average weight is about 900 pounds. One-half cent per pound above the prices ruling during the season of 1887, and in all probability Texas would not have shown one failure in the cattle business.

On Wednesday of last week, Texas cattle sold up to 3 cents. . . . Exporters, including fancy Herefords and Shorthorns (in Missouri and Illinois), weighing 1,700 to 1,970 pounds brought $5.20 to $5.70. These cattle were raised and fattened on high priced land, where more feed is required to sustain life than would fatten a steer in Texas.

We argue, first, that Texas beef is produced at the lowest cost, and that if a margin of profit remains we have it. Our advantages are, briefly, cheap lands and a mild climate. . . . Our disadvantage of raising scrub cattle is fast passing away, consequently it is only reasonable to suppose that Texas will continue to produce cattle cheaper than other states.

It is estimated that 400 pounds of cottonseed, with hay, sorghum or cactus for roughness, will fatten a steer. This should not cost more than $2, delivered on any ranch in Southwest Texas.

There will probably be 20,000 head of cattle fed on cactus and cottonseed in Southwest Texas during the next four months. This is just a beginning, and next year there will be at least 100,000 fed.

Historical and Biographical Record of the Cattle Industry

Brownwood Cotton Oil Company Feedyard, Brownwood, Texas, proved that "aged" Longhorns can be fattened. Ca. 1890.

Library of Congress

Eufaula Cotton Oil Company, Eufaula, Indian Territory, was built in 1898 and fed cattle until 1933. Ca. 1907.

Chapter 2

Early Feedyards In The Southwest

At the turn of the 20th century, a number of feedyards were feeding 1,000 cattle or more.

As Texas, Oklahoma and New Mexico ushered in the 20th Century, several large feedyards were in existence – most operated by the cottonseed oil mills. They were feeding cattle primarily to dispose of their by-products – meal and hulls – which in most cases had been hauled away and dumped into low places.

The first cottonseed oil mills in the U.S. were built in the 1850s, just before the Civil War. By 1895, the U.S. had 205 oil mills, of which Texas had 84 and Indian Territory (I.T.) had several. (Indian Territory became the State of Oklahoma in 1907.) Several of these feedyards had capacity for 1,000 head or more.

Feedyards in Indian Territory

One such mill was the Eufaula Cotton Oil Company, Eufaula, I.T., which opened in 1898 and continued feeding (as Osage Cotton Oil Company after 1914) about 1,000 cattle until 1933. Robert L. Simpson, Jr. of Eufaula worked there in 1932-33. "My father was president in 1933 when the price of meal dropped to $7 per ton," he recalls. "We just couldn't make it, so my father shut down the mill and the feedyard."

In 1913, the Interstate Cotton-Seed Crushers Association, headquartered in Dallas, reported that "for years the Durant Cotton Oil Company of Durant, Oklahoma, had fed thousands of cattle at their mill."

And the Chickasha Cottonseed Oil Company at Chickasha, Oklahoma, had feeding pens early in this century which some local citizens believed to be "the largest in the world." But according to J. Harper Thomas, a cattleman at Chickasha, "The oil mill didn't actually operate the feedyard – it furnished the feed. Cattlemen rented adjacent pens and cared for their own cattle."

As a young boy, Thomas worked there on weekends, helping his father who was one of the largest and most steady customers during the 1920s and 30s. "Another large customer was L. W. Long of Rush Springs – he was sort of the dean of feeders and was well respected," says Thomas. "At any one time, the Chickasha pens had several thousand head. Our rations were simple – meal and hulls, corn and molasses, which were fed by hand from mule-drawn wagons."

First Feedyards in Texas

In Texas, oil mills feeding cattle numbered in the scores. One of the first was the Brownwood Cotton Oil Company at Brownwood, which started about 1890 with a capacity of about 1,000 head. Another was the Planters Oil Company, which operated in the Brazos Bottom near Hearne from 1900 to 1927.

Another early feeder was the Simmons Cotton Oil Mills, established by four brothers from Oklahoma – Will, Bob, George and T.B. Simmons. By the late 1920s, they had mills (most with feedyards) in Lubbock, Rotan, Sweetwater, Quanah and Childress, Texas; also in Roswell and Loving, New Mexico.

T. B. Simmons, Jr. started at Quanah, has fed cattle almost continuously for 62 years.

Tom B. Simmons, Jr. of Lubbock started in the business in 1930 and fed cattle nearly every year until 1992. His first exposure to the feeding business came at the oil mill pens at Quanah. The pens there, with capacity for 7,000 head, were all of wire, sufficient for keeping the cattle confined normally, but scant restraint against the big stampedes that started after occasional lightning storms. The feedlot wasn't big by today's standards, concedes Simmons, "But 7,000 head is a whale of a bunch when they're scattered in every garden and cotton patch in the county."

Most feedyards of the day were located on draws – the more direct the drainage the better. At the 6,000-head feedlot in Lubbock, "We drained straight into the Yellow House Fork and never thought anything of it," says Simmons. "Everybody tried to get as close to a stream as they could."

Drainage engineering may have changed considerably since the 1930s, but one thing in the cattle feeding business – the risk factor – seems to remain pretty constant: Tom Simmons bought his first cattle in 1931 – two car loads – from the Lee Bivins Estate in Amarillo for $5.50 per cwt. He shipped them to Quanah for a "quick" 300-days or so of feed, and then to Chicago. They brought the young entrepreneur $3.60 per cwt.!

The oil mill pens were, in their day, the giants of the industry. Roy Weeman, the Swift buyer from 1912 to 1961, said if he or other buyers had need of a 'cattle on

Courtesy Irwin Munn

H. B. Johnson feeding pens, near Chickasha, Oklahoma. Ca. 1910. Other Johnsons – brothers E.B., T.T. and Ira Johnson – owned feedyards at Minco and Norman, where they fed several hundred head per year..

feed report,' "We'd just drop by a few of the oil mill pens and ask around. They had most of the cattle, and they knew as much about the business in the 1920s and 30s as anybody." Weeman, in addition to buying cattle for Swift in Fort Worth, placed Swift cattle in Texas and New Mexico Feedyards – in more than 20 feedyards at times.

Though the oil mill feedyards numbered in the scores and though their capacities were considered "large," there were years when fewer than 50,000 head were fed in the state – less than some individual yards finish in the 1990s.

Most of the cattle the mills fed were owned either by the oil mills or by the ranchers. Simmons Oil Mills in Lubbock, for instance, charged ranchers 1 cent per day "pen rent" and tacked most of their costs into the price of the feeds they sold.

Since the oil mill pens were the first feedyards to regularly offer pen space to outsiders, they must be considered the forerunner of the modern custom feedyard, even though their primary goal was marketing by-products, not making beef.

Courtesy Albert Carter

Since the objective was to dispose of by-products, Otis Gin Feedyard at Loving, New Mexico fed cattle for long periods. This steer weighed 2,480 pounds and sold for 9 cents per pound in the late 1930s.

"Meal Evil"

Though they thrived, the oil mill pens were not without their problems. Among the worst was "meal evil" or "meal blindness."

Meal evil was discovered shortly after cattlemen first started feeding full rations of meal and hulls. After a few months on the ration, cattle went blind. (See Chapter 5.)

Tom Simmons ran across more than his share of meal evil during his early exposure to meal-and-hull cattle. "We had no trucks," he says, "so we had to drive our cattle down the road from the pens to the railroad. A bunch of blind cattle weren't easy to drive!"

The word on how to prevent meal evil seemed slow to spread. As early as 1915, ***The Cattleman*** offered its congratulations to W. A. Wilhelm of Quanah for selling 90 head of 1,370-pound steers in Oklahoma City for $8.60 per cwt. "They'd been fed on meal and hulls and after 100 days," reported the magazine, "some of the steers were beginning to suffer with blindness from the effects of the meal. . . . And right here is where Mr. Wilhelm displayed his ability as a feeder. He reduced the cottonseed meal and hulls, and added to the ration alfalfa, maize, meal and Tarkio. Those suffering from the effects of the meal quickly recovered, and all continued to gain in weight. . . . "

Another feeder who early learned the remedy to meal blindness, without knowing the cause of it, was Sid J. Webb, owner of the Riverside Cotton Oil Co. in Fort Worth. He fed alfalfa with his cottonseed products as early as 1925, and told ***The Cattleman*** that year his lack of meal evil problems had something to do with the green color of alfalfa.

Meanwhile, some farmer-feeders were feeding cattle on a scale that would be respectable even by today's standards.

Col. C. C. Slaughter of Dallas was feeding 50,000 range cattle in 1895. "Meal is the finest feed in the world," he said, "because it gets cattle ready for market in from 80 to 90 days."

Texas and Southwestern Cattle Raisers Foundation

Native steers on feed at the G.W. Morris ranch, 20 miles east of Waco, Texas, during the winter of 1892. Note feed wagon pulled by four mules.

A neighboring rancher, Winfield Scott of Fort Worth, fed 2,100 steers on meal and hulls in 1895 and said, "There is no finer feed on the earth. I have tried it in the pens and in the pasture, and I believe the best way to feed is for a man to put his meal and hull troughs right in his pasture, so that cattle can get some grass once in a while if they want it."

Another well-known rancher, W. T. Waggoner of Decatur, fed 2,000 head on meal and hulls in 1884. "The finest lot of steers my eye ever rested on," he called them. "We fed them 90-100 days, and sold them . . . weighing over 1,300 pounds . . . for $4.25 and $4.50 per cwt."

"Texas To Become No. 1"

These glowing reports led to optimism by some observers, such as Greenlief W. Simpson of the Fort Worth Packing Company and Fort Worth Stock Yards. In an 1895 speech to the 22-year-old Cattle Raisers Association of Texas, he took a five-year look into the future. "Six hundred thousand cattle could be fed in Texas on cottonseed oil meal," he projected. "Whoever lives to see 1900 will see more cattle fed in Texas than in any other country in the world."

Later in 1895, Simpson, who traveled frequently from his native Boston to Chicago and Fort Worth, boasted in a local stock journal that Texas ranked as "the largest cattle feeding state in the nation."

About 1894, R. E. Gatewood of Cleburne, Texas advertised his feeding operation: "Manufacturer of High Grade Beef. Prime Beef A Specialty." From 15 silos on his ranch, Gatewood fed 2,000 to 4,000 head a year, some in drylot and some on pasture, and marketed them at about 1,300 pounds in Kansas City.

Courtesy W. G. McDonald

D. L. McDonald Feedyard, built in 1912, probably was the first feedyard in Deaf Smith County, Texas. These light heifers sold in Kansas City, May 3, 1914, for $8.75 per cwt. – then a record. McDonald (left) also dug the first irrigation well in Deaf Smith County, 1910.

"Once I sold corn-fed steers weighing 1,255 pounds at $2.55 per cwt.," he said. "But they cost me only $17 each and were fed 17-cent corn." Between 1894 and 1914, his books showed returns ranging from a loss of $12 per head to a profit of $20 per head.

D. L. McDonald had a baby beef operation near Hereford, Texas prior to 1914, where he fed kafir silage, alfalfa, milo and cottonseed meal. On May 3, 1914, he sold 100 Hereford yearling heifers averaging 686 pounds in Kansas City for $8.75 per cwt. – a record price at the time for short yearling heifers from the Texas Panhandle.

Tradition had it that finished beef under three years lacked flavor and there was little export demand for young cattle. The Swensons of the SMS Ranch at Stamford set out to disprove that, however.

Frank Stewart Hastings, the SMS manager in the early 1900s, credited T. F. B. (Tom) Sotham with giving him the idea of pushing range-bred calves to be matured as finished beeves in the Cornbelt at an age not exceeding 20 months.

Hastings experimented. He put weaned steer calves on a maintenance ration of cake and hulls, sorghum hay and black-strap molasses and sold them a year later at ages of 18 to 22 months. Said he: "We then went to the Cornbelt with the broad

Texas and Southwestern Catle Raisers Foundation

Col. Ike T. Pryor, for whom La Pryor, Texas was named, financed farmers to build silos and took the payback in silage at $4 per ton fed to his cattle; or he would sell cattle to the farmers. Ca. 1914.

Many silos were built in the early 20th Century. Hale County, Texas had over 100 silos in 1916.

statement that well-bred Texas calves were good enough for any Cornbelt feedlot."

Many Texas ranchers looked to the Cornbelt feeders as "the market" for feeder cattle. They toured the Midwest, promoted their cattle and invited Cornbelt feeders to "come see the plentiful supplies of quality cattle in Texas."

The stigma against "Texans" (Texas Longhorns), which were slow to finish, was beginning to wear off in the early 1900s. So Midwest feeders responded favorably. They now could buy quality Hereford, Durham and Shorthorn calves or yearlings and ship by rail; thus, many Texas cattle became grand champions at the International Livestock Exposition in Chicago, which was the prestige show.

But some Texans remained convinced there was a future for cattle feeding locally.

John C. Burns, head of Animal Husbandry Department, A. & M. College of Texas, wrote in 1915: "In this new era of beef production, silage is playing a most important part and will continue to do so more and more. . . . Texas (will become) noted for its finished cattle, as it has long been noted for its feeders."

Out-of-state feeders were beginning to agree. C. O. Keiser moved from Iowa to Canyon, Texas about 1911 and brought Cornbelt feeding expertise with him. In 1915, he sold 70 head of steers in Kansas City for $9.15 per cwt., believed at the time to be the highest ever paid on the open market for Texas cattle.

Then Came Silos

Another Cornbelt feeder, R. R. Hammond of Illinois, saw the Texas potential and bought 34,000 acres of land near Miami, Texas, where he built 14 large pit silos and started feeding cattle. The day has passed, he said in 1915, when Texas cattlemen will be content to breed good cattle and allow others to put them through the feedlot. Thousands of silos are found on the ranches of this State and, with the increased production of feedstuffs, "Texas is surely destined to become the greatest producer of fed cattle."

Confirming this prognosis, Z. E. Black, a writer for ***Country Gentleman*** magazine, traveled through Hale County, Texas in 1916 and was impressed with the silos. From one windmill tower, he wrote that he "could see with the naked eye 27 of the county's 100 silos. . . . It is claimed that the 800-ton silo of H. E. McCabe, Hale County, is the largest in the world." And he wrote that Texas had put up 2,000 silos in 1913.

The idea of feeding silage soon reached South Texas. In 1914, Col. Ike T. Pryor, who for many years had ranked high among the "Cattle Kings" of Texas, owned 100,000 acres in Zavala County, about 100 miles southwest of San Antonio. (Pryor served as president of the Cattle Raisers of Texas in 1906-09 and president of the American National Live Stock Association in 1917-18.) After he allowed several thousand acres to be colonized, the community became know as La Pryor. Promoted as part of the "Winter Garden," La Pryor became known for its cabbage, lettuce and Bermuda onions.

But the "Winter Garden" bubble soon burst and Col. Pryor sought a new tack to bolster the local economy. So he called a meeting of farmers and offered to furnish the money to any responsible farmer to build one or more silos. Pryor would take his pay in silage fed to his cattle at $4 per ton. Or he would sell cattle to the farmers, plus the cottonseed meal at cost, and take his pay from the sale of cattle in Fort Worth.

As a result, 12 modern silos were built and this was called "the dawn of a new and prosperous era for the farmers of La Pryor."

Silos also were being built in Oklahoma. In 1912, Charles N. Bancroft of Rickford (Blaine County) built two metal silos, with a capacity of 200 tons each, and that winter fed Kafir corn silage to 230 steers.

Other prominent Oklahoma feeders in the early part of the century were the Johnson brothers – E. B., T. T. and Ira – who fed several hundred head each year at Minco and Norman.

In 1916, T. T. Johnson won the first annual Southwest Live Stock Show at Oklahoma City with a carload of Hereford steers that weighed 1,398 pounds and sold for $9.35 per cwt., the highest price ever paid until then on the Oklahoma market. On the regular market at that time, most steers were bringing $90 to $110 per head.

M. T. Johnson of Amarillo tells about his grandfather, E. B., who was impressed with the grass in the Texas Panhandle and started buying land in 1905. "His idea was to (1) breed and start the cattle on his ranch at Abilene, (2) graze them as yearlings in the Texas Panhandle, at Borger and Alanreed, (3) feed them in Oklahoma, and (4) ship them by train from Oklahoma City to market in Kansas City."

Panhandle-Plains Historical Museum

Members of a Baby Beef Club show their calves on Main Street in Hereford, Texas. Ca. 1920.

The El Paso Times

J. C. Peyton of El Paso started a feedyard in 1917 on the Rio Grande River, later expanded it to 17,000 head and built a packing plant (shown in foreground). In 1962, when President John Kennedy agreed to give this Chamizal area back to Mexico, Peyton found his feedyard and packing plant in Old Mexico. (Juarez, Mexico is shown in upper left corner of photo.)

Even earlier, E. B. Johnson started feeding cattle on a sizeable scale. About 1890, he and his father, Montford Johnson, obtained a contract to supply Fort Reno with fresh dressed beef for 8 cents per pound. Because there was no refrigeration, they would move about 100 head at a time from their feedlots to the west side of Fort Reno and herd them there until time for slaughter. They butchered twice a week, usually 10 head at a time, and delivered the beef fresh. E. B.'s beef and hog feedlots – the hogs to clean up after the cattle – expanded to include a whole section.

First Baby Beef Club

In the early 1900s, bigger was considered better when it came to fed animals. But cattle numbers were dropping rapidly – from 50.5 million in 1900 to 35.8 million in 1914 – and a few cattlemen thought it more efficient to feed young animals than the conventional two-, three- and four-year olds. J. E. Boog-Scott, a large rancher-feeder at Coleman, Texas, concluded the best way to advance this idea – at the same time to train future cattlemen – was through boys. So he and a couple of neighboring ranchers, with the help of a professor at Texas A&M, in 1912 organized the first Boys Baby Beef Club, which was a forerunner to 4-H Clubs. Young Sam Matthews was the first president.

"Membership is open for boys between eight and 18 years of age," said Matthews in a 1914 speech to the Cattle Raisers Association of Texas. "The boy

Texas and Southwestern Cattle Raisers Foundation

Automatic feed car in the Texas Panhandle, 1915, delivered feed out the side at the rate of 50,000 pounds per hour. Location unknown.

must feed his animal at least 120 days, keep an account of everything he feeds it and then show it."

"The objective," continued Matthews, "is start your boy out while he is young, educate him for the business, and he will be worthwhile." Also, he reasoned that better cattlemen would help rebuild the nation's cattle herd.

But the Baby Beef movement ran into some opposition from "humanitarians, armchair theorists and politicians." In 1916, Congressman F. A. Britton of Illinois introduced a bill to prohibit the slaughter of all beef cattle under two years of age. Any offender would be punished by a fine of not less than $100 for each live cattle or carcass offered for sale.

The congressman's contention was that the production of baby beef would reduce production and increase the cost to consumers. But that legislation failed.

World War I, More Beef Needed

World War I began in 1914 and the U.S. entered in 1917. J. C. Peyton, an El Paso, Texas butcher who had two small markets and a small slaughtering plant, had a contract to supply beef to the army at Fort Bliss. But the demand was outstripping his ability to buy finished beef. So in 1917, he moved to the south edge of El Paso, on the bank of the Rio Grande River, and built a new packing plant plus feeding pens for 3,000 head of cattle. The feedyard grew through the years until it ultimately had a capacity for 17,000 head.

But a hitch developed for Peyton Packing Co. and Peyton Feedyard. The problem dated back to 1848 when Mexico lost the war with the United States and the boundary was established at the Rio Grande. But rivers change. And for a century, Mexico claimed that about 630 acres in southern El Paso legally belonged to Mexico.

So in 1962, President John Kennedy met with President Adolfo Lopez-Mateos and agreed to give the Chamizal area back to Mexico. And suddenly, Peyton discovered that their packing plant and feedyard were in Mexico! (They were compensated, however, and relocated the packing plant on the eastern edge of El Paso and the feedyard 40 miles downriver at Tornillo.)

World War I brought on more concerns, in addition to the "packer trust" issue, which soon reached a boiling point. (See Chapter 4.)

National Cottonseed Products Association

To dispose of meal and hulls, Otis Gin & Warehouse Co. (oil mill) of Loving, New Mexico operated this 1,000-head feedyard, leased from Rich R. Carter, from the late 1920s through the 1930s. Ca. 1930.

Desperate cattlemen proposed the building of their own packing plants and discussed the raising of capital to finance them. A few municipalities did build abattoirs, such as Paris, Beaumont and Austin.

"Excessive profit" by beef retailers also was being challenged. According to government figures as of July 15, 1918, the price of sirloin steak in Chicago (typical of several cities surveyed) was 37.7 cents per pound, when fed steers were averaging $17.62 per cwt. The retail profit was 3.7 cents per pound, which cattlemen considered excessive.

President Woodrow Wilson in 1917 appealed to the patriotism of American farmers, as the U.S. entered the War, and asked them to increase food production. Also, Secretary of Agriculture D. F. Houston requested Congress to fix a minimum and maximum price on food products.

Cattlemen Oppose Price Controls

This prompted Texas Gov. James E. Ferguson to appoint a committee of cattlemen – Col. Ike T. Pryor, president of the American National Live Stock Association; James Callan, president of the Cattle Raisers Association of Texas; and Col. R. J. Kleberg of the King Ranch – to call on Food Administrator Herbert C. Hoover. On May 31, 1917, they prevailed on Mr. Hoover to not fix either a maximum or minimum price on meat producing animals. "It would demoralize the business," they said, "frighten many producers out of the business entirely, and could not be successfully carried out."

Next came "Beefless Meals" and "Meatless Days," imposed by Food Administrator Hoover, sometimes called "Food Dictator." The government's rationale was to make available to our army and allies 75,000,000 pounds of meat each week, compared to the pre-war normal of 15,000,000 pounds.

Texas and Southwestern Cattle Raisers Foundation

"A well bred, well formed and well finished steer," fed by S. E. McKnight of Sonora. It sold in Fort Worth, June 1916, weighing 1,750 pounds, for $10 per cwt.

On top of all this, concern was growing about the change in the dietary habits of American consumers. In August 1918, just before the War ended in November, James Callan, president of the Cattle Raisers Association of Texas, wrote:

"For 20 years, in one form or another, directly or indirectly, there has been a defamation of beef and a campaign for substitutes. . . . Yet we have never spent a dollar in all these years to offset the publicity urging substitution. . . . The day will come when every factor in meat production – breeding, development, finish and distribution – will join in clean, vigorous and intelligent publicity to keep the mouth of consumption knocking at the door of production."

FIRST FEEDYARD IN DEAF SMITH COUNTY

When James B. McAndrews moved from Iowa to Hereford in 1956 and bought Western Feedyards, some residents considered that to be the first feedyard in Deaf Smith County. (Western was built a few years earlier by geologist Homer Cogswell, but it was empty in 1956. McAndrews expanded it from about 1,000-head capacity to 2,500 – until the railroad came through and took most of the pens.)

Records at the Deaf Smith County Museum show, however, that D. L. McDonald might have built the first feedyard, 45 years before McAndrews arrived. McDonald, a land developer and real estate promoter from Ohio, drilled the first irrigation well in the county, in 1910, and later drilled 50 more. He also was the first president of the Hereford Chamber of Commerce, in 1918.

Marketing his abundant irrigated crops was difficult, so McDonald built the feedyard as a means of selling feed, an approach he had observed in the Midwest. From his "Baby Beef Factory," one set of calves, weighing 686 pounds in 1914, topped the Kansas City market at $8.75 per cwt.

Paul Engler arrived from Nebraska in 1960, built (with eight other investors) the Hereford Feed Yards . . . and the boom was underway! By 1972, the Southwestern Public Service annual survey showed 574,108 fed cattle marketed and Deaf Smith was hailed as "The No. 1 Cattle Feeding County in the U.S." The number continued to increase until 1989 when 885,563 were marketed.

Neuhoff Packing Co.

Neuhoff Feedlots, built south of Dallas in 1962, had 30,000-head capacity. Auger in center ran to five self-feeders, each 1,900 feet long, so one man could feed 8,000 head in three hours. Auger also loaded trucks for filling outlying feeders.

Chapter 3

BIG FEEDYARDS ARRIVE

While several big feedyards operated during the 1930s, they grew in numbers and size during the 1950s.

In the three decades that followed World War I, cattle feeding in the Southwest expanded very little. Numbers on feed in Texas, for example, reached 300,000 in 1932, then declined and didn't rebound until well after World War II.

Nevertheless, the winds of change caused a number of "outsiders" to move into the area, as usually occurs following a war, and there was an ever-growing demand for Yankee-type beef. As a result, food stores and packers were having some difficulty in the 1920s filling the new demand.

Roy Weeman worked for Swift, 1912-61, and became head buyer and manager of cattle feeding. Ca. 1970.

Early Packer Feeding

Roy Weeman, the Swift buyer, said the production chain wasn't too quick to pick up the signal: "We couldn't find anybody to feed all the cattle we needed." Thus, packers began to feed their own cattle. Two of the most active were Peyton Packing Co. and Swift & Co.

Peyton, in addition to their feedyard at El Paso (Chapter 2), opened a 5,000-head feedyard at Littlefield about 1932. "This was considered a big feedyard at the time and the first in Texas with alley bunks," says O. J. Barron, Jr. of Spur and Lubbock, who got his first feeding experience with Peyton in 1932 and who became president of TCFA in 1976. "Until then, it took two men to feed – one to drive the wagon team and the other to open gates so we could get to the feed troughs."

Barron's father was vice president of Peyton, in charge of procurement, and young Barron's main job was weighing in bundles, cattle and hogs. "This yard operated at Littlefield until 1942," he adds, "when it was closed, under state orders, following an outbreak of anthrax."

Paul D. Bennett, who would become mayor of Littlefield, started work at the Peyton feedyard in 1937 and served as manager until 1942, when he left for the army. Explaining the anthrax outbreak, he said, "After bringing in cattle from the Kansas City and Oklahoma City stockyards, we started losing two or three fats a day. After losing about 60 head, the problem was determined to be anthrax and Peyton shipped the remaining 12,000 head or so to El Paso. Four years later, in 1946, Swift rented the yard and brought in 4,000 head. Soon, these cattle also started losing weight and dying, so they were shipped out and the yard closed for good."

About 1920, Swift had a try at building and operating its own feedyard. Impetus was L. S. Swift, Jr., who came down from Denver to manage the Texas operation. He brought with him J. D. McCrimmons, who had been feeding Swift cattle in Colorado.

They built a feedyard with about 3,000-head capacity, which Roy Weeman remembered as the most modern in the state at the time. Among its facilities was the first electric feed mill in Texas. Unfortunately, the feedyard was built about two miles north of the Fort Worth Stockyards, and within a couple of years complaints by neighbors caused Swift to close it.

That was Swift's only feedyard in Texas, but the company kept some cattle on feed to fill the demand for fed cattle. In 1923, they began feeding at the Riverside Cotton Oil Mill feed pens, six miles northwest of the Fort Worth Stockyards. Also, Swift fed cattle in the old horse and mule barn at the Stockyards and later placed cattle with farmer-feeders around the state.

In 1928, Weeman purchased 1,000 Choice heifers from Dalhart for Swift and had them fed by George Pridemore of Coleman. In 1935, he began placing Swift cattle in Davidson Feed Pens at Pecos, a yard that grew to 20,000-head capacity and continued to be operated by Paul C. Davidson until his death in 1989. During the 1930s and 1940s, Swift also fed cattle with Fred Snyder of Lubbock, Ray Ayres of Slaton, Henry Lawson of Lubbock and W. C. Weatherly of Tornillo.

Beginning in the 1940s, Swift got serious about buying fed cattle. Chain stores were becoming more and more important in the meat market, and the chain buyers knew what kind of beef they wanted. "That was the first time we ever had anybody come into our cooler looking for beef," said Weeman. "If they didn't see what they wanted, it didn't pain them a bit to turn around and walk out. About the second time that happened, packers made darn sure they had what the chains wanted the next time."

"TEXAS BRED AND FED"

It became a slogan throughout the state, a rally call for all of Texas agriculture, the motto for bumper stickers: "Texas Bred and Fed."

The Texas Breeder-Feeder Association was organized in September 1929 in Midland, where it had been conceived by Paul Vickers, secretary of the Midland Chamber of Commerce. Frank P. Holland, editor-publisher of ***Farm & Ranch*** magazine in Dallas, was elected president and became an active leader-promoter. The Breeder-Feeder Movement, as it became known, had wide support from cattlemen and farm organizations, the Extension Service, universities, bankers, railroads, chambers of commerce and others.

The original purpose was to "put livestock on the farms of the Black Belt of Texas, where soil fertility was being rapidly depleted by the one-crop system (cotton)." But the movement spread to every section of the state where feed was being raised. Local affiliates were formed, like the Northeast Panhandle Breeder-Feeder Association at Canadian and the West Texas Feeder-Breeder Exchange at San Angelo.

The Exchange at San Angelo, a not-for-profit organization, had the stated purpose, "to promote this movement in any way it can . . . to act in an advisory capacity and as a clearing house . . . to bring farmers and ranchmen together who are interested in feeding on a cooperative basis." It offered three contracts. The No. 1 agreement for example, provided for "the farmer to feed the livestock and after they are sold to pay the ranchman for his original weight of stock, the market price less 1 1/2 cents per pound."

The movement enjoyed remarkable success for a few years. In January 1932, A.L. Smith, Extension livestock specialist, reported that "Texas has nearly twice as many cattle feeders as the previous year and about 50 percent more cattle on feed." In April 1933, A.L. Ward, education director for the Texas and Oklahoma Cottonseed Crushers Association, wrote: "It has given to Texans a new dream of the great possibilities of livestock feeding in our own section."

The Breeder-Feeder Movement, and farmer feeding, continued to be promoted after World War II and until commercial feedyards began to appear in the 1950s.

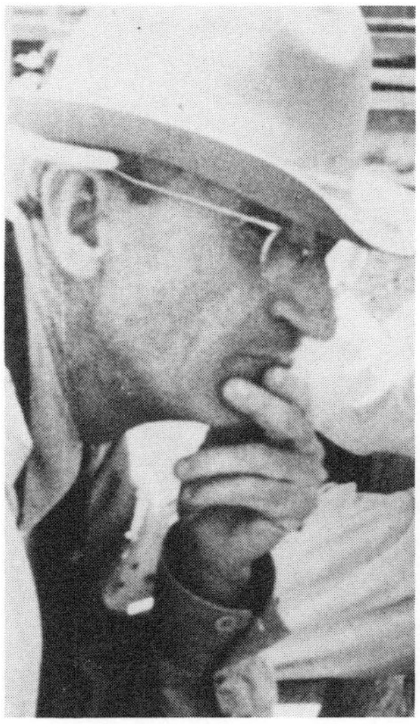

Texas Cottonseed Crushers Association

Fred Snyder of Lubbock, with feedyards at Levelland, Whiteface and Coble Switch, 1930-41, made money right through "The Great Depression." Feedmill (above right) was at Whiteface, 1937.

What the packers wanted in the 1940s was a 550- to 600-pound carcass, which left out the 1,300-and 1,400-pound "Prime" animals of the day. Their demand that the beef be well-marbled also axed the grass-feds.

During the 1940s, Swift settled into a cattle production routine which persisted for several years. Calves were placed on wheat pasture which had been leased from farmers in the Texas Panhandle. When the wheat ran out in the spring, they were taken to feedlots.

Pete Buske of Friona was among those who helped place and oversee Swift cattle on wheat pasture in the Plains, running as many as 2,500 head for the company some winters. L. M. Price of Stratford was another, along with Roy Whooten of Dalhart.

In 1942, Swift placed 5,000 head with Milton Rheinholdt of Dalhart. "Until then, there weren't 100 head of drylot cattle in the county," said Weeman. However, after Rheinholdt had worked with many area farmers to get the cattle fed that winter, Gooch Elevator at Dalhart built feed pens. They opened in 1945 and Swift fed with them for several years.

Swift, along with Peyton and Armour, were doing all they could during the war years to secure fed animals. Few Texans were really interested in feeding cattle without a "partner" to share the risk. Packers were doing it because they had to, because that's what their customers – the growing grocery chains – demanded.

Weeman says the advent of the chain stores did more to tempt the consumer to accept, then demand, top-quality beef than any other part of the industry. When the consumers demanded it, the industry had no choice but to start producing it.

Depression Era Feeding, Between Wars

As the Great Depression of the 1930s approached, several sizeable feedyards were operating on the South Plains of Texas. Two were at Brownfield – the D. H. Snyder feed pens and the Baricora Development Co. feedyard. The latter, owned by newspaper magnet William Randolph Hearst, had 10,000 head on feed in 1937.

Only a few miles up the road, another large feeding operation was run by Fred Snyder of Lubbock. In 1920, Snyder and his partner, D. C. Reed, an Austin investor, leased 150,000 acres of pasture west of Lubbock. They'd purchased 5,000 cows just before the 1929 crash, thus entered the Depression heavily in debt.

Snyder was intent on working his way out, and in his search for a passage his eyes fell upon the depressed farm markets. Silage around Lubbock could be bought

Texas and Southwestern Cattle Raisers Foundation

H. W. Stanton operated a 3,000-head feedyard along the Yellow House Canyon of Lubbock, 1931-46, which was recognized by Ripley's "Believe It or Not" for having the world's longest water trough.

for $1.50 per ton, cattle for as little as 3 and 4 cents per pound. He and Reed decided to put the two together, and in 1930 opened feedyards at Levelland, Whiteface and Coble Switch. The three yards put the partners up among the biggest rancher-feeders in Texas. They finished 10,000 head of cattle and about that many sheep in 1937. The feedyard profits (they fed primarily cattle from their ranches and others they purchased for the feedlot) helped them get out of debt during times which put most people in debt, according the Snyder's son, R. L. Snyder of Clayton, New Mexico. The elder Snyder operated the feedyards until 1941, when he liquidated both his ranch and feedyard holdings because of failing health.

Shortly after Snyder started, a Lubbock farmer and feed miller, H. W. Stanton, built another big yard, this one curving along the Yellow House Canyon which was to be engulfed by a growing Lubbock. Stanton had a sizable farm in the area and after the lots were built in 1933 channeled much of his farm produce through the lot in the form of ground bundles and silage. The feedyard was notable not only for its size, but for its facilities.

Stanton's daughter, Mrs. Mary Alyce Fehleison of Sisterdale, recalls her father believed that cattle always needed a supply of clean, running water. To provide it, he had the feedyard pens equipped with what was billed by Ripley's "Believe It or Not" as the world's longest water trough – 2,230 feet running the full length of the pens. One well ran continuously into the trough, night and day, until 1946, when it went dry and the feedlot was closed.

The size of the Stanton operation also attracted attention. Mrs. Fehleison recalls that on one cattle-buying trip into East Texas, the Lubbock feeder purchased 10,000 head of cattle. All of the cattle had to be transported back to the Plains on rail cars, and railroad authorities were unable to provide Stanton enough cattle cars and still take care of normal business in the region.

At the time Stanton died, shortly after Pearl Harbor, he had 12,000 head of cattle and several thousand head of sheep in the feedyard.

The successful ventures of Snyder and Stanton graded a lot of roads which would be paved and traveled by future generations of Plains feeders. They had gone "big time" and had showed that Plains-fed beef could compete with Cornbelt animals: Texas-fed cattle were finally topping Midwestern markets.

Right behind them came Claude Keeton, who had moved to Lubbock about 1923

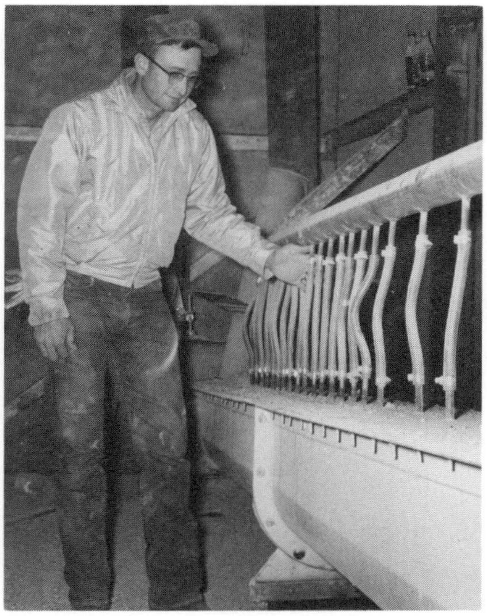

Photos by Charles E. Ball

J. O. Parker of Happy, Texas (right) began feeding in the 1930s and by 1966 he and his son, Foster, had 4,000 head on feed. Foster Parker (above) shows equipment for soaking and reconstituting grain sorghum, which increased feed efficiency 10-15 percent.

from his native Fannin County. He initially made his way with a backyard slaughtering operation and during the 1930s established Keeton Packing Co.

Shortly after he had the packing operation going, he bought pens which had been built in 1923 by another Lubbock packer-feeder, Alec McDonald. Keeton used the pens, located near the Lubbock Cotton Oil Mill's feedyard, for finishing his own cattle primarily. But he also finished beeves for other packers, according to his son, Sidney Steve Keeton of Lubbock.

During the late 1930s and 1940s, Claude Keeton also leased the yard at the Slaton Cotton Oil Mill. The Keeton feedyards stayed in operation until 1965.

In 1937, Homer Hill of Hart, Texas began a 42-year career of feeding cattle, starting with a 200-head feedyard which by the beginning of World War II was expanded to 3,000-head capacity. Later, in 1962, it was replaced with a newer and larger feedyard, which he operated until his death in 1979 and which still is operated by the family as Hill Land & Cattle Co.

In the meantime, about 1950, Homer Hill and Byron Bennett built a yard at Plainview to be known as Bennett-Hill Feedyard. When J.C. Roderick bought out Bennett, it became Hill-Roderick Feedyard, until Roderick bought out Hill in 1971.

J. O. Parker was another of the early feeders on the Plains. His estimate of cattle feeding was such that in 1932, when he was offered the choice of either continuing his collegiate football career at New Mexico A&M College or overseeing that college's cattle feeding experiments he dropped football and fed cattle.

After working as a county agent in New Mexico for several years, Parker purchased two sections of wheat ground near Happy, Texas and immediately began feeding 50 to 150 head of cattle each winter. Over the coming years, he added land and penspace until, in 1966, he and his son, Foster, had 4,000 head on feed, primarily their own cattle.

One of the first feedyards to bring year-around feeding to Texas was the Sudan Livestock and Feeding Co. at Sudan. Ben H. Davidson and Anton Rieder opened the

Photo by Charles E. Ball

BOUNTY HUNTERS

STARLINGS CAUSE US MUCH DAMAGE THROUGH FEED EATEN AND DAMAGED. THEY ALSO ARE VERY UN-SANITARY. AS AN INCENTIVE TO REDUCE THEIR NUMBERS, WE ARE OFFERING UNTIL FURTHER NOTICE A BOUNTY OF

15¢ A BIRD

FOR ALL STARLINGS KILLED WITHIN A TWO-MILE RADIUS OF SUDAN AND DELIVERED TO US.

SUDAN LIVESTOCK

AND

FEEDING COMPANY

Early custom feeder Ben H. Davidson (right) opened Sudan Livestock and Feeding Co. in 1940. Disgusted with starlings, he posted sign (center) in 1960, offering a bounty. Only about 100 birds were brought in. Young men, first attracted by the offer, found the cost of shotgun shells made the offer economically prohibitive. Ben's son, Jim Davidson (left), became manager 1967-79.

feedyard during the fall of 1940 with 2,000 head. When Davidson sold out to Rieder during the latter part of World War II, they were feeding 5,000 head.

Ben's son, Jim Davidson, bought back into the company in 1954, and recalls there were only 500 head of cattle on feed at the time: "We were just existing. The market had broken something awful, from $34 per cwt. in 1952 to $16 in 1953."

Relief for Davidson came in 1956 when the Red River Land and Cattle Co., a division of the Matador Ranch, put 2,000 calves into the Sudan yard. In 1957-58, substantial expansion was begun and, by 1972, the lot had a one-time capacity of 25,000 head. In the 1990s, Sudan Livestock and Feeding Co. is operating under the same name but under different ownership – CDH Feeders, Inc.

In the 1940s, large commercial yards were established with some regularity, some destined for success, others doomed to early failure – prey of economic and climatic

Photo by Charles E. Ball

Norman Moser of DeKalb had one of the first feedyards in East Texas, about 1938-62, until switching to custom feedyards in West Texas. He was first chairman of the TSCRA Cattle Feeders Division.

conditions which were pushing the hub of the state's feeding industry toward the Plains. Still, many farmers established feedlots in other parts of the state and prior to 1960 there were a number of commercial lots established in East and South Texas.

Among the more successful feedyards in East Texas was that of Norman and Otto Moser of DeKalb, who started feeding cattle in the late 1930s to provide a market for the products of their farm.

Over the next 20 years, they built the feedlot to 6,000-head capacity, feeding silage and by-products of cotton produced on their own and neighboring farms. Over the years following World War II, however, the region drifted away from crops, which were replaced with tame grasses.

By the time the feedyard boom hit West Texas, "We were having to import all our feeds," says Norman Moser, "so we decided it would be cheaper to ship our cattle out there than to ship everything back here." They sold the feedyard in 1962 and started feeding cattle in the commercial pens of the Plains.

Following World War II

Further west, different motivations had produced the same sort of idea for Ray Smyth of Aledo, near Fort Worth. Smyth had started out as a farmer-feeder and, beginning during the depression, had built a thriving feed business.

In 1950, Smyth put in a feedyard near his mill with the original idea of its serving as a market outlet for his feedstuffs. The pens held 10,000 head of cattle – impressive and unique enough that for several years Neiman-Marcus included the Aledo Feedyard on the tour it afforded visiting foreign dignitaries. (See page 153.) During the ensuing years, the feedyard, almost without competition for a while, paid off well through feeding contracts with neighboring ranchers and Fort Worth packing plants.

Death blow to the Aledo yard came during the 1960s when the boom in feeding on the High Plains first drew packers' attention, then their plants. By 1977, Smyth phased the yard out of cattle finishing and began feeding in custom feedyards on the Plains.

Another early Texas feeder was E. George Luckey, owner of the T. Cross Cattle Co., west of San Antonio. His privately-owned yard, which later became known as Luckey Custom Feedlot, Inc., was started in 1949 with 6,000 head capacity and grew to 15,000 by 1975 when he died. Senator Luckey, who had served in the California State Senate, had fed cattle in California during the 1930s and had become one of the largest feeders in the Imperial Valley during the 1940s.

Wanting to return to his native Texas and feed cattle, he found a 1,500-acre site in Bexar County in 1948 that was suitable for a farming operation and feedyard. "When I bought the tract, I was going to use it to collect cattle for shipment to the West Coast," Luckey said. "But the potential for feeding cattle in the San Antonio area changed my mind."

Union Stock Yards, San Antonio

George Luckey returned to Texas from California, where he had been a state senator and a large feeder, to build a feedyard near San Antonio, 1949.

Another mid-century feedyard in South Texas was the Green Valley Cattle Co. at San Marcos, built in 1950 and owned by Jim Cummings of San Marcos and I. C. Little of Happy. Cummings was an inventor – with 62 U.S. patents over 50 years, the most lucrative of which was the bulldozer – and Little made his money as a pipeline contractor.

It was a model feedyard, which gained much publicity, with 12,000-head capacity and a number of unique features. Such as: a Williamson mill with steam flakers, believed to be the first in Texas; a multi-directional spray room, which replaced the dipping vat; and it was built in connection with an auction.

But there were drawbacks. Such as: being built too close to town, so odor and flies caused complaints; runoff into the river, causing protests by environmentalists; and poor drainage, because it was built on flat land in a river bend.

Courtesy Ernest Cummings

Green Valley Cattle Co. at San Marcos, Texas, built in 1950 by I.C. Little (left) and Jim Cummings (right in left photo), an inventor, had several engineering innovations. Covered troughs (right photo) accommodated hay above and grain below.

Caddo Wright, the first manager who is in the feed business in Lubbock, tells about one time the river overflowed: "We had one hill in the feedyard, but the 12,000 cattle soon filled that and we ended up swimming many cattle out."

Moody Taylor, co-manager in 1953, showed two pens of Mexican crossbred steers that weighed 1,400 to 1,500 pounds. "Last month, the owner turned down $20 per cwt.," said Taylor. "Last week, he turned down $16; and this week the best bid is $14." A week later, Taylor reported that the owner had committed suicide by jumping out of a hotel window in Fort Worth.

Since then, when the cattle market slipped into the red, that story has served as a reminder: This isn't the first time we've had a bad market.

Green Valley Cattle Company sold in 1962 to Bob Cooper, a native of Iowa, who completely closed it about 1975.

Back on the High Plains, Walter Lasley, a farmer near Stratford, began feeding cattle in 1953. He started with 150 head and expanded until, in 1967, he and his sons had capacity for about 2,000 head. That year, to accommodate neighbors who wanted in on what Lasley was doing, he built a feedmill, added penspace to hold 10,000 head of cattle and formed Walter Lasley & Sons, Inc. Managed by his son, Walter E. Lasley, the yard in 1992 has room for 18,000 head.

In the early 1950s, Durward W. Lewter, a county agent at Big Spring, had learned how to feed milo to 4-H Club steers and win blue ribbons across the nation. Believing the technique would work commercially, he secured the financial backing of oilman Clint Murchison of Dallas and in 1955 built the largest commercial feedyard that had been built in West Texas. Lewter thus brought into focus the profit potential of large-scale cattle feeding – and subsequently the hazards inherent in that business.

During a taped interview for the Southwest Collection of Texas Tech University, Lewter pointed out one of the problems feeders faced during the 1950s: Packers had to be sold on milo-fed cattle. "Grain sorghum-fed cattle didn't fit into the eastern and northern markets," he said. "I finally sold a load to A&P in the East and was told I had a better product than Cornbelt feeders, because they over-fattened their cattle, making them wasty and less attractive to the consumer. A&P wanted more of my milo-fed steers."

"Packers and the chain stores had built an image for corn as THE feed to fatten cattle," Lewter added. But after winning several carcass shows with milo-fed steers, he helped change that image. He fed mainly company-owned cattle to High Choice and Low Prime grades and expanded the yard until, in 1960, he had nearly 30,000 head of cattle on feed. During the market drop of the early 1960s, hard times fell on

Durward Lewter, with backing of Dallas oilman Clint Murchison, in 1955 built a 30,000-head feedyard at Lubbock – the largest commercial feedyard that had been built in Texas. A former county agent, Lewter had a reputation for feeding fancy cattle.

Photos by Charles E. Ball

Lewter as they did on smaller feeders. (In protest of low prices, the National Farmers Organization called a strike in September 1962, and thousands of Midwestern farmers withheld their cattle and hogs from market.)

A legal controversy developed between Lewter and Murchison and, in 1962, the feedyard was sold by the court to E. C. Crofoot and his son, Jay, of Kansas, who operated it as Lubbock Feed Lots, Inc.

Harold D. King, Muleshoe, fed exclusively for Swift from 1960 to 1975, then made King Feed Yard a custom feedyard. Ca. 1980.

In 1960, Harold D. King opened King Feedyard at Muleshoe, and for the next 15 years fed exclusively for Swift & Company. According to King's son, Max, "We had a 10-year contract, then a 5-year contract, that prohibited us from feeding for anyone else. Swift bought the cattle and feed, formulated the rations and told us what to feed. Their on-premises man sorted the cattle to fill in their kill (at their Fort Worth plant). Only after those contracts expired did we start custom feeding for other customers." (This feedyard continues to operate in 1992 but under other ownership and another name, Wes Tex Feed Yards.)

In 1962, when Roy Weeman retired from Swift, where he had worked 49 years, including two decades as head cattle buyer, his long-time friend Harold King prevailed on him to come to work for King Feedyard as fat cattle salesman. This he did for the next 10 years, all the time commuting from Fort Worth to Muleshoe.

In the Rio Grande Valley, Jack Scoggins, who has operated feedyards at San Benito and Rio Grande City since 1959, tells about the first big feedyard in that area:

Walter Lasley (right), a Stratford farmer, started feeding in 1953. He's visiting here, 1980, with Max Carpenter of American Cyanamid, one of approximately 500 Industry Associates who have supported TCFA through the years.

Photo by Charles E. Ball

E. C. Crofoot, a well-known feeder in Kansas, bought the ailing Lewter Feedyard in 1966 and renamed it Lubbock Feed Lots.

"The Iverson brothers – Pete, Leon and Ike – moved in from the North and bought a citrus orchard. Being of Nordic descent, they felt the need for natural fertilizer; so in the early 1940s they put in a 3,000-head feedyard north of Alamo – primarily for the manure. They fed only Herefords, because they said crossbreds – admittedly more plentiful and cheaper in South Texas – tore down their fences. The Iverson Bros. Feedyard expanded to about 4,500-head capacity before closing in 1960."

East Texas Mud

In East Texas, A. G. Morton of Kilgore started a 6,000-head feedyard in 1958, under the name of Texas Feed and Cattle Co. Over the next few years, several lots were built within 50 miles – at Carthage and Nacogdoches to the south and at Daingerfield to the north.

By 1963, when Morton closed the Texas Feed and Cattle Co. yard, most of the other large feedyards in East Texas had also been shut down. Although they had plenty of cattle nearby, the feed supply was too little and the mud problem was too much.

A group of Houston businessmen opened the North Houston Feedyard in 1961. With capacity for 7,000 head, the feedyard was situated across from Armour's Houston packing plant, and most of its cattle were marketed to Armour. Plagued with mud problems and high transportation costs for feed that had to be shipped in, the feedyard was closed after only five years of operation.

Perhaps the largest feedyard to operate outside the Plains in the 1950s and 1960s was the Milford Farms Feedyard between Milford and Hillsboro. Opened in 1962 by Neuhoff Packing Co. of Dallas, and later called Neuhoff Feedlots, it was designed to provide the top-quality cattle Neuhoff had trouble finding in those days.

Peak capacity was 30,000 head and most of the cattle fed were owned and processed by Neuhoff. With the growing number of feedyards in Texas, Neuhoff found in the early 1970s it could buy all the fed cattle needed. This, plus the same losses wracking the rest of the industry, caused the company to study the idea of going into custom feeding. But it was barred from that enterprise by a Packers and

Oklahoma Farmer - Stockman

Ralph H. Mock, a pioneer feeder at Altus, Okla., started feeding in 1920. His sons, Paul and John, took over in 1952, expanded the feedyard to 5,000 and continue operating it into the 1990s. Ca. 1964.

Stockyards Administration prohibition against packers owning custom feedyards. So the feedyard was closed in 1974.

Other sizable yards opening for business in Texas during the 1950s were Dumas Cattle Feeders, Inc., Dumas; Moody Farms, Pampa; Easley Feed Yard, Hereford; Carter-Kirchhoff Feed Yard, Plainview; Western Feed Yard, Hereford; Cluck Cattle Co., Hart; Lykes Bros. Feed Yard, Inc., Edroy; Cattle Feeders, Inc., Goliad; Nacogdoches Cattle Co., Nacogdoches; Keeling & Sons Feedlot, Leona; Triplett & Pringle Cattle Co., Bovina; Bills & Ellis Feedlot, Fabens; McCloy Feedlots, Morse; Shepard Feedlot, Hale Center; Anton Feedlot, Anton; and League Ranch, Benjamin.

Oklahoma Pioneer Feeders

Courtesy George S. Barnes

Sid Barnes of Hobart, Oklahoma built a 500-head feedyard in 1943, fed with a one-horse feed buggy, and expanded to 3,500 in 1955. Ca. 1947.

In Oklahoma, a pioneer feeder was Ralph H. Mock. At the age of 98 and living in Altus, he recalls first starting as a farmer-feeder in 1920, upon returning from World War I. "Through the 1920s and 30s," he remembers, "we would run 200 to 300 cattle on stalks, then move them into a pen with hogs where we fed Hegari bundles. Our source of water was a creek that ran through the feeding pen. The creek also made for good drainage, but I guess the environmentalists would frown on that today."

Mock's sons, Paul and John, took over the feeding operation in 1952, expanded to 5,000 head capacity and merged with the (Don) Kizziar Feedyard in 1970 to become Cattle Management Systems, which still operates in 1992.

In nearby Hobart, Oklahoma, Sid Barnes built a 500-head feedyard in 1943 with J. D. Simms, who Barnes bought out in 1955 when the capacity was 3,500. George S. Barnes, born in 1944, claims to have "grown up in the feedyard" and became manager upon his father's death in 1972.

George remembers that in the early years they mixed feed on the floor with a scoop shovel and hauled it to the cattle with a feed buggy drawn by one Percheron horse. Also, he remembers that in the 1950s and 60s they couldn't get packer buyers out to the feedyard, "except the one night a year that we had a steak dinner. So we would ship the cattle to Oklahoma City, Fort Worth, Kansas City or St. Joseph and take what the packers would give us," he continues. "Except occasionally my dad did bring some cattle back (when the bid was too low), just to make a point, even though it wasn't always economical to be shipping fat cattle back and forth."

Ray Schoeb built a 3,000-head feedyard at Cherokee, Oklahoma in 1949, to utilize dehydrated products, and fed in self-feeders (left). Schoeb (above, center) was assisted by sons Robert (left) and Jerry (right). Ca. 1957.

Sid Barnes got his early experience in cattle feeding while working for the Chickasha Cotton Oil Company, beginning in the early 1920s. His job was to ride the caboose of the train with the cattle and, when the train stopped, to check the cattle and get up any down animals. The next day, he would watch them sell – a practice of the day by most feeders who didn't trust the commission companies – and bring the check home.

In 1932, feeling the pressure of the on-coming depression, Kansas automobile dealer Roy Schoeb moved his family to Cherokee, Oklahoma, where he started farming, then built an alfalfa dehydrating-pelleting plant, then in 1949 built a feedyard to better utilize and market his farm products. "We had little interest in the Democrat's farm programs," Schoeb said in 1955. He was a staunch conservative, who later became a Republican state senator.

Schoeb's sons, Robert and Jerry, joined him in operating the feedyard where they specialized in pelleted feeds – alfalfa, grain and forage sorghums. This permitted them to use self-feeders, which were refilled every five days. From an initial capacity of 3,000 head, the feedyard grew to 7,000 head when it was sold in 1979.

Old time employees and traders on the Oklahoma City Stockyards remember the Oad Colbert Feedyard, which started in the late 1930s just north of the stockyards. Luther Eoff worked there after World War II and says the capacity was 10,000 to 12,000 head at that time. "Most of the customers were commission companies," he says. "On horseback, we used to drive cattle across the river to the feedyard and later drive them back across the river to the packing plants. The packers required us to stand the cattle overnight without feed and then shrink them 3 percent."

Fred Ridley, Sr. who also worked in the stockyards – for his father who operated National Commission Company – also fed cattle with Colbert.

When to Hold 'Em, When to Fold 'Em

Ridley tells about buying 1,500 cattle in Kissimmee, Florida in 1932 and feeding them at the Cushing Cotton Oil Mill feeding pens in Cushing, Oklahoma: "The 500-pound yearlings cost 7 cents per pound. We fed them $3-per-ton hulls and $7-per-ton meal, until they reached 900 pounds, and the cattle still lost money. They sold for 5 cents per pound.

"A couple years later," he continues, "I bought 325 calves in Odessa, Texas. Paid $7 per head for 300, and the man gave me 25 of the lighter ones, which weighed about 150 pounds. These calves had a lot of age, but it was during the drouth of 1934 and they weren't doing too well – living off burned pear."

Topping his own hard-luck story, Ridley says things were even worse when he was a young boy, right after World War I: "My dad, R. Courts Ridley, and Roy Williams of Amarillo in 1919 bought about 500 steers from a rancher, just north of Amarillo, for $95 per head. And they'd been better off if every one had laid down and died.

"The next day, Roy was bid $105 per head but turned it down, asking $115. The market broke so they sent the cattle to the Osage (northeastern Oklahoma) to graze for one year. Then they sent the cattle to a feedyard in Oklahoma City and fed them through the winter. When bid only 3.5 cents per pound, they sent the cattle back to grass in 1921. Finally, they shipped the cattle, weighing about 1,200 pounds, to Kansas City where they sold for 4 cents per pound or $48 per head."

About 1941, the Oklahoma Livestock Marketing Association (OLMA), a cooperative owned by National Livestock Credit Corporation, got into the feedyard business in Oklahoma City. "With World War II coming on, the old Colbert Feedyard was sitting there nearly empty, so we leased part of it from Mrs. Colbert,"

In 1953, Henry C. Hitch (left) and H.C. (Ladd) Hitch built their first feedyard at Guymon, Oklahoma, the first of four yards. By 1988, Hitch Enterprises had become the eighth largest feeding company in the United States. Ca. 1962.

Lee Merrill (seated) started feeding cattle in Clovis in 1932 – cattle he sold for $4.25 per cwt. Then he built this 2,000-head feedyard in 1942. Also shown in this 1948 photo is Ted Waldhauser (standing), who came from St. Paul and bought his first feedyard in Clovis in the 1940s.

Livestock Weekly

Ted Waldhauser later, in 1961, bought the Clovis Feedyard with Wendell Keenan. Ca. 1951.

related Harley Custer, OLMA manager from 1934 to 1979. "Our plan was to offer a new service – cattle feeding – to our customers. So we operated this as National Feedlots until 1960, when we built a modern new feedyard west of Yukon, Oklahoma." The new feedyard had a capacity of 10,000 and operated until 1982.

Further west, in the Oklahoma Panhandle, M. H. Freeman in 1941 built his first feedyard, consisting of four pens with 2,500-head capacity. Freeman lived in Texhoma, Oklahoma, but his feedyard was on the southwest edge of Texhoma, one-half mile into Texas. After 10 years, he closed it down and built a 5,000-head feedyard 11 miles to the southeast, again in Texas. Here he installed a Williamson feedmill, manufactured in Montebello, California, which he believed to be the first in the Panhandle. It's still in operation, 41 years later.

M. H. Freeman was known as a plunger and a gambler, but he claimed to never have brought a weaned calf home that lost money. Before he died in 1959, the feedyard was taken over by his son, Jack M. Freeman, who continues to operate it in 1992.

Bigger Feedyards To Be Built

Still bigger feedyards were yet to be built in the Oklahoma Panhandle, by the Hitch Ranch, which had its beginnings about 1885. Henry C. Hitch and his son, H. C. (Ladd) Hitch, drilled their first irrigation well in 1947, several more during the drouth of the early 1950s, and by 1958 had 19 wells irrigating about 4,000 acres of grain sorghum, alfalfa and wheat.

In 1952, the Hitches sent some of their own ranch cattle to a custom feedyard in Arizona and later observed that the feedyard was purchasing its grain from the Oklahoma and Texas Panhandles. "They bought our cattle and they bought our grain," says Ladd, "and the only asset they had was a little better winter weather for feeding cattle."

So in 1953, the Hitches constructed their first four feeding pens. This was the beginning of Henry C. Hitch Feedlot, which expanded to 1,000-head capacity in 1955, to 5,000 by 1958 and ultimately to 45,000. In 1965, Ladd led a company with nine investors to build Texas County Feedyard at Guymon. Two years later, another Hitch-led group built Master Feeders (now called Hitch Feeders I), capacity 60,000.

By 1988, Hitch Enterprises had become the eighth largest feeding company in the United States, with Paul H. Hitch, a fourth generation rancher-feeder, as CEO.

New Mexico Pioneer Feeders

Early on in New Mexico, the feedyards tended to concentrate around Clovis and in the Pecos Valley below Roswell, where feeds were plentiful.

One of the first feeders at Clovis was Cash Ramey, a real estate developer who owned two small feedyards (300 to 400 head each) in the early 1930s. Wendell Keenan, who lived in Clovis most of his life and later in Russellville, Arkansas, describes Ramey's feedyard design: "Troughs ran out from old boxcars where the feed was stored. He had no mill – a couple of guys would come, custom grind bundles, sack the feed and store it in the boxcars. Thus, the workers could walk right out of the boxcars into the troughs with the feed sacks.

"The way he mixed feed was creative, too," continues Keenan. "I remember seeing two men with high-top overshoes pouring molasses onto the feed in the trough and mixing it with their feet."

Another early feeder, Lee Merrill, came to Clovis in 1932 and started feeding cattle (in addition to ranching), before building his own 2,000-head feedyard in 1942. At age 80, he tells of the time he and Cash Ramey accompanied their six cars of fed cattle to Kansas City for sale. "Ramey had four cars of fancy Herefords that were fat and weighing about 1,000 pounds. I had two cars of plain red steers that weighed 1,150 pounds. On the way up, we made a deal: Ramey would furnish the cigars while I would keep the stove fired in the caboose.

Fancy Herefords Brought $3.75 Per Cwt.

"The market wasn't too good in 1934," continues Merrill. "His fancy Herefords brought $3.75 per cwt. while my plain steers brought $4.25. Ramey was furious! And on the way home, he claimed that he couldn't afford cigars."

Z A. McCasland, who has fed cattle in the Clovis area since 1939, adds this story about Cash Ramey's "marketing prowess." W. D. Mack, who ran the yard for Ramey, always carried a sling shot in his hip pocket. When a poor animal would come to the fore, Mack would shoot him, sending him to the back of the pen. Thus, the poor animals were trained to move to the background when a man arrived – so the packer buyer never got a good look at them.

Ted Waldhauser, an order buyer for Wertheimer Cattle Co. out of St. Paul, came to Clovis in 1938 and in the 1940s bought a small feedyard, where he fed cattle for Swift. In 1961, he and Wendell Keenan bought Clovis Feedyard, which was built during World War II by Ben Davidson for feeding hogs. Davidson, a former order buyer for Coast Packing Co., Los Angeles, had a contract with Cannon Air Force Base to dispose of their garbage, which he fed to hogs. While continuing to live in Clovis, Davidson later bought Sudan Livestock and Feeding Co. at Sudan, Texas. (Davidson's senior partner in the Sudan feedyard was Anton Rieder, with Coast Packing Co.)

Following World War II, Clovis Feedyard sold a couple times and in 1961 was purchased by C. B. Erdwurm and Wendell Keenan. In 1975, Erdwurm bought out Keenan and into the 1990s the yard was owned by his stepson, Jim Lewter.

Among other feedyards in eastern New Mexico dating back to the 1950s were Mesa Livestock, Inc. at Clovis, built by J. D. Hance and Hal Bogle, and since 1979 owned by Harvey Tindell; and Green Valley Feedyard at Tucumcari. Green Valley, with a capacity of 4,000, was operated by Don Flemming until 1960 when it was sold to Worley Mills – owned by brothers Clarence and Olan Worley and their three sons – who closed it in 1983.

Bogle Farms Feedlot at Dexter, N.M., was built in 1937 with fenceline bunks – an innovation in those days. By 1973, Hal C. Bogle and his son, Bill, operated three feedyards– at Dexter, Clovis and Chandler, Arizona – with a combined capacity of 30,000 head.

To the Southwest of Clovis – in the Pecos Valley between Roswell and Carlsbad – a dozen or more feedyards emerged during the 1960s and 1970s. One was even earlier – the Otis Gin and Warehouse Co. feed pens at Loving, which fed 600 to 700 head a year, from about 1920 to 1934. Otis owned a cotton oil mill and leased the pens from Rich R. Carter Farms as an outlet for their meal and hulls.

Hal G. Bogle moved to Dexter from Tennessee in 1919, for health reasons, and later founded Bogle Farms, Inc. In 1932, he started feeding cattle and in 1937 built a feedyard with fenceline bunks – an innovation in those days – for 2,000 cattle. Bill Bogle became a partner with his father and by 1970 they had over 12,000 head on feed – all Bogle-owned cattle.

When Hal Bogle died in 1973, they were operating three feedyards – at Dexter, Clovis and Chandler, Arizona – with a combined capacity of 30,000 head.

Robert O. Anderson, internationally-known oil and gas businessman at Roswell, bought a small feedyard in 1960 from Sacra Brothers, renamed it Diamond A Feedyard and expanded it to 25,000-head capacity by 1963, then to 45,000 by 1971. In the meantime, it was leased to Janss Cattle Industries of California (1963-69), then sold in 1971 to Glover Packing Co. and renamed Pecos Valley Feedyard. But hard times fell on Glover, because (a) the Packers and Stockyards Administration ruled in 1972 that a packing plant could not operate a custom feedyard and (b) they were caught in the wreck of 1973 with 40,000 head of company-owned cattle. Glover closed the feedyard in 1974. Later, it was bought and operated (1978-81) by Far Tex Feeders, Inc., a group of 14 investors led by Dean Stallings.

Among other feedyard operators in the Pecos Valley between 1960 and 1985 were V. R. Barnett, Royce Lankford, Richard Harshey and Bill Langenegger, all at Hagerman; Carl Nicholas and H. C. Berry at Dexter; Rolla Hinkle and Albert Watson at Roswell; and Gil Moutray near Carlsbad. None of these were left as finishing yards in 1992; instead, large dairies had taken their places.

Early Beef Promotion

Early in the 20th Century, cattle industry leaders were concerned about the image of their product and were advocating more beef promotion.

"For 20 years, in one form or another, directly or by inference, there has been a defamation of beef and a campaign for substitution," wrote James Callan, president of the Cattle Raisers Association of Texas, in 1918. "It began with patent medicines, has been taken up to some extent by the medical fraternity, has been vigorously pushed by cereal manufacturers, and at the present time our clever friends in the dairy industry are making a broad campaign and spending a vast sum of money in publicity looking to the substitution of dairy products for meat.

"Our attention has been called to the fact that we have never spent a dollar in all these years to offset the publicity urging substitution," he continued, "and that the day would come when every factor in meat production, breeding, development, finish and distribution would join in clean, vigorous and intelligent publicity to keep the mouth of consumption knocking at the door of production."

Two years later, at the 1920 convention of the American National Live Stock Association, Frank D. Tomson ridiculed the headlines of that day, such as: "Eat Less Beef," "Meat Hurts the Kidneys" and "Meat Causes Rheumatism." "The remedy lies in a well-organized and educational campaign," he declared, "which will include the aid of reliable and skilled culinary experts and that will be at least nation-wide in scope."

Early on, some physicians came to the defense of the meat industry. For example, Dr. W. H. Lipman wrote in the 1921 San Francisco Journal of Commerce: "The authorities are quite generally agreed that the people who eat the most meat have by far the most ambition and initiative and greater resistance to disease than those who eat little meat."

He was reinforced by Dr. William H. Porter, dean of the New York Post Graduate Medical School, who that same year wrote a book, ***Eating to Live Long***, in which he said: "Vegetarians as a class are almost invariably ailing in some respect. I have yet to see a strict vegetarian who did not have something the matter with him, even if it was only a chronic grouch."

A couple years earlier, in 1919, packer Thomas E. Wilson made a speech to the Kansas Livestock Association which generally is credited as "the seeds" for the National Live Stock and Meat Board. He suggested that producers meet with packers to develop a "proper organization" to represent a united industry. But when the producers arrived for the meeting they discovered that the packers already had a plan for a 33-member board – comprised of 11 packer representatives, 11 market representatives and 11 producers – which aggravated the suspicions that producers already had about packers. The "fair representation" issue was ultimately comprised, however – after the Packers and Stockyards Act clamped down on packers – and the National Live Stock and Meat Board was founded in 1922.

The Meat Board grew steadily through the 1930s and 40s, with an initial collection of 10 cents per car on cattle (5 cents paid by producers and 5 cents by packers). This was gradually increased to 2 cents per head on cattle, 2/3 cent on hogs and 2/5 cent on sheep, which by the 1950s was bringing in $1.1 to $1.3 million per year for red meat promotion.

That wasn't enough to please cattlemen, however, who were hurting from the drought and price break of the early 1950s. They wanted to promote "beef," not "meat." So they started the independent National Beef Council (NBC), under the leadership of Jay Taylor, Amarillo cattleman who was president of the American National Cattlemen's Association (1954-55), former chairman of the Meat Board (1951-53) and later a member of TCFA. Their plan was to collect a checkoff (8 cents per head) through the market agencies, the same collection procedures being used by the Meat Board.

But the Meat Board objected, as did the Farm Bureau, Farmers Union and National Grange. Also, the USDA ruled that the Packers and Stockyards Act prohibited such a collection at "posted markets," unless the producer agrees to it each time he ships cattle in.

NBC officers cried "discrimination," because the Meat Board was using market agencies to collect, and in 1962 filed a suit against the Meat Board, which later was dropped. They also alleged that the Meat Board was dominated by big packers and chain stores.

Tempers flared! And leaders said both organizations were endangered. In response, the Meat Board reorganized and a Beef Promotion Committee was appointed to handle beef funds. That took the steam out of the National Beef Council.

About the same time, John Armstrong (then an Alabama cattleman, later of Kingsville, Texas) was asked to arbitrate and attempt to consolidate. He did and the result was the Beef Industry Council of the Meat Board, formed in 1963.

A Texas Beef Council was organized in 1954, the purpose being "to increase beef consumption in Texas." Membership grew to about 2,000 producers and associates, with local chairmen in 212 of the 254 Texas counties. But the TBC dissolved in 1963, as did the NBC, both blaming their demise on the USDA ruling that market agencies could not collect the checkoff.

In 1969, TCFA made a stab at promoting "Texas Grain-Fed Beef," in cooperation with the Texas Department of Agriculture. Two years later, TCFA sponsored a "Beefiesta" in Dallas and Atlanta – in cooperation with the BIC of the Meat Board – which was the beginning of TCFA's involvement in national beef promotion.

In 1974, TCFA proposed a national legislated checkoff. Fourteen years and three referendums later, in 1988, it became effective – the $1-per-head checkoff that we have today.

Texas and Southwestern Cattle Raisers Foundation

Two or three times a day, cattle in custom feedyards are fed a scientifically balanced ration, formulated by a consulting nutritionist.

Chapter 4

THE ENTRY OF BIG GOVERNMENT

Concerns of cattlemen have changed surprisingly little during this century.

During the first half of the Twentieth Century, cattlemen began to feel the helping hand of government, in the form of government programs, government agencies and government regulations. Some called it "the heavy hand," because some of the schemes were forced upon them "in the public interest." But most of the programs lasted and impacted all cattle feeders to come.

The Packer Trust

One hundred years ago, cattlemen were upset with packers – big packers who, cattlemen alleged, were unethically and unlawfully colluding and taking advantage of stockmen. And through the years, the allegations changed little.

As early as 1888, a Select Committee of the United States Senate was appointed to conduct hearings on the transportation and sale of meat products. Cattle prices had dropped 50 percent – to $3 per cwt. for Texas steers – so cattlemen from across the country eagerly testified in hopes of relief. Afterwards, Senator G. G. Vest of Missouri summed up the feelings: "We know that today five firms in the City of Chicago control the price of beef throughout the whole United States. There is not a butcher from Boston to Charleston but who feels the effect of this combination."

At nearly every meeting of the pioneer associations, misconduct by the big packers, along with "excessive freight rates" by the big railroads, were topics of protest.

At the 1905 convention of the Cattle Raisers Association of Texas, for example, A.B. Robertson of Colorado City charged in a fiery speech: "There is a small coterie of men seated in palatial offices in the City of Chicago, who daily place the price arbitrarily on every hoof of livestock that is sold. . . . They are pursuing the community of interest plan that seems to be wonderfully effective in stifling competition.

"Surely, there is a remedy for this situation," Robertson concluded, "and in my opinion the remedy lies in the control of the packing houses by the laws of our country."

W. W. Turney of El Paso, president of the Cattle Raisers at that time (1905), added fuel to the fire. "On the question of the existence or the non-existence of a packer trust," he said, "We have got to continue to deal with the beef buyers and the beef packers as long as time lasts. . . . If you cannot do without the packers, you want to have confidence in the man who buys. . . . I am sorry that the people (big packers) who

consume 98 percent of our product cannot represent things openly and fairly back to us."

"In my opinion, there is a combination existing against your interests," he said, "and it exists in violation of the Sherman Antitrust Act. If it does exist, there should be honesty and integrity enough to stamp it out."

The discontent increased through the years and, as World War I began in 1914, the price of beef increased significantly. This prompted the Department of Agriculture to appoint a national committee to investigate the cause of high beef prices.

In 1916, ***The Cattleman*** magazine noted, "The market for fat cattle during the past few years has been distinguished by the absence of competition, and producers and feeders are clamoring for relief."

This feeling prompted the American National Live Stock Association (predecessor to ANCA and NCA), meeting in El Paso in 1916, to appoint a Market Committee "to ascertain and enforce a line of justice and fairness between the producers of livestock and the few men who buy their product."

This was followed by congressional investigations . . . hearings . . . meetings with packers . . . proposals that cattlemen build their own packing plants . . . a suit by the Justice Department . . . and eventually an investigation by the Federal Trade Commission (FTC) which lasted over two years.

Power of the Big Five

So in 1918, the FTC issued its report, which said in brief: "The power of the Big Five packers in the United States (Swift, Armour, Morris, Cudahy and Wilson) has been and is being unfairly and illegally used to: Manipulate livestock markets. . . . Restrict interstate and international supplies of foods. . . . Control the prices of dressed meats and other foods. . . . Defraud both the producers and food consumers. . . . Crush effective competition. . . . Secure special privileges from railroads, stockyard companies and municipalities. . . . Profiteer."

An antitrust suit by the Justice Department prompted the packers to propose a compromise, and in 1919 the Justice Department agreed. This was the famous Consent Decree, in which the packers admitted no guilt but consented to the following:

1. To sell all their holdings in public stockyards.
2. To sell all their interests in stockyard railroads and terminals.
3. To sell all their interests in market newspapers.
4. To dispose of all their interests in public cold storage warehouses.
5. To forever disassociate themselves with the retail meat business.
6. To forever disassociate themselves with . . . wholesale groceries.
7. To forever abandon the use of their branch houses, route cars and auto trucks.
8. To submit perpetually to the jurisdiction of the United States District Court under an injunction forbidding conspiracy or monopoly.

At last, cattlemen had won their case! Or so they thought. But a year later, packers were not living up to their agreement, and the battle renewed.

In 1920, John B. Kendrick, president of the American National Live Stock Association (ANLSA), gave an eloquent speech to the convention in Spokane, Washington. He praised the work of the ANLSA Market Committee, which had been formed four years earlier in El Paso and which had been successful in getting the attention of congress and in getting action from the Federal Trade Commission. But that wasn't enough – he insisted that "legislation is needed."

"No man need fear the results of the legislation that we ask," he said. "We want to prevent interference. We want to tear down artificial restraints; we want to abolish arbitrary restrictions. We want a free American market, a market which every day shall register the value of a consignment of livestock as accurately as the scales register its weight."

Packers and Stockyards Act

So, Kendrick, a native Texan who also was a United States Senator from Wyoming – probably the only man to hold such a high public office while serving as president of a cattlemen's association – introduced legislation which led to the Packers and Stockyards Act of 1921. (A House compromise version finally was adopted, rather than the Senate bill.)

However, the sentiment to regulate packers was not unanimous. At the 1920 convention of the Cattle Raisers Association of Texas, President W. W. Turney accused the packers of false advertising regarding their "low profits." Ed C. Lasater of Falfurrias joined Turney in speaking for regulation. But Judge William Poindexter of Cleburne called it "un-American and un-democratic." He was joined in opposition by well-known cattlemen like F. S. Hastings, manager of the Swenson ranches at Stamford, Robert J. Kleberg of the King Ranch and M. Sansom of Fort Worth.

Support for regulation came from a new organization, the American Farm Bureau Federation (AFBF). In May 1921, twenty-five senators met in the Washington office of AFBF and agreed to block the adjournment of Congress until the Packers and Stockyards Act and other farm bills were passed.

The legislation did pass and on August 15, 1921 President Warren G. Harding signed the Packers and Stockyards Act. It gave the Secretary of Agriculture supervision over the packers, stockyards, commission men, traders, buyers and sellers in the stockyards.

Immediately after the Act went into force, on November 1, 1921, the packers contested its constitutionality. But on May 1, 1922, the United States Supreme Court upheld all provisions of the Act and it has been in force ever since, although with numerous amendments or changes in regulations.

One immediate result of the Act was the introduction of a new practice – direct marketing. Since packers had to give up ownership of their stockyards and commission companies, packer buyers simply bypassed them and began buying direct from feeders. Some producers charged that packers were doing this in order to pay lower prices and avoid regulations at the stockyards. Sure enough, direct buying increased through the decades and no doubt contributed to the downfall of major stockyards.

Meat Inspection for 100 Years

Meat inspection – of some kind and to some degree – has been practiced throughout most of recorded history. Both the slaughter and marketing of meat were inspected in ancient Athens and Rome. Jewish sacrificial animals had to be perfect enough to be eaten by a priest, so in time the rabbi's authority was stretched to include all meat in Jewish communities.

During the 12th Century, laws passed in England, France and Germany banning the sale of meat from diseased animals, and some type of inspection system has been in existence in these countries ever since.

In the United States, there was little concern about meat safety during colonial times, because sanitation was an individual responsibility. It worked, since most buyers knew the butcher and oftentimes knew the producer. But as transportation developed, butchers and packers began selling meat farther away and individual responsibility became meaningless.

As John H. McCoy explained it in his book, ***Livestock and Meat Marketing*** (1979): The prevailing doctrine was "caveat emptor" or "let the buyer beware."

In the late 1800s, the U.S. had developed quite an export trade to Europe. But opposition developed among the foreign buyers – for economic, political and alleged sanitary reasons.

So in 1890, in an attempt to regain European confidence in American beef, Congress passed the first meat inspection statute, which applied only to exported meat. It didn't accomplish its goal, however – Europeans still refused to buy American meat – so the act was strengthened in 1891 and 1895.

Cattlemen welcomed the move. At an 1895 meeting of the Cattle Raisers Association of Texas, Robert J. Kleberg offered the following resolution, which was adopted: "Resolved: That the Cattlemen's Association thoroughly, heartily and unevasively, without any reservation, commend and support the present rigid method of cattle and meat inspection. . . ."

In 1906, Upton Sinclair wrote a novel, ***The Jungle***, based on conditions in Chicago packing plants, which obtained much publicity and attracted the attention of President Theodore Roosevelt. So Roosevelt met with Sinclair and decided to push for strict regulations.

The result was the Federal Meat Inspection Act of 1906, which called for mandatory inspection – both antemortem and postmortem – of all meat and meat products moving in interstate commerce. It also established sanitary standards within slaughter and processing facilities. At this time, it was determined that meat inspection was "in the public interest," so it became a tax-supported federal program.

The next major change came in 1967, when the Wholesome Meat Act required that all state plants were to have standards at least equal to those of federal requirements. Prior to that, there existed a hodgepodge of state and municipal regulations which varied from adequate inspection to none at all. The final result was that the number of meat plants with inspection expanded greatly – from 163 in 1907 to 22,000 federal and state plants in 1980.

Meat Grading: A Common Language

As commerce developed in the beef business, the need for standards or "common language" in marketing became more apparent.

In 1916, Congress had established a national Livestock Market News Service, so that producers and others might have an unbiased daily market report. But the lack of uniform standards was causing much misunderstanding, confusion, suspicion and, in some cases, material losses.

Two government agencies – the U.S. Shipping Board and the Veterans Bureau hospitals – requested standards, so the USDA prepared tentative ones in 1923 and published them in 1924 in a bulletin, "Market Classes and Grades of Dressed Beef." These were the first grade standards for meat issued by the Department of Agriculture.

The editor of ***Breeders Gazette***, Alvin H. Sanders, became enthused about the possibilities and wrote that "beef from well fattened cattle of excellent conformation and quality should be marketed as Blue Ribbon Beef." Other editors, managers of livestock shows, retailer and livestock associations jumped on the bandwagon.

So in 1926, the Secretary of Agriculture promulgated the earlier beef grade standards as the official U.S. Standards for Market Classes and Grades of Carcass Beef.

On July 22-23, 1926, a meeting of 250 prominent cattle breeders and feeders from across the country met in Kansas City to discuss the advisability of grading and stamping the two highest grades of beef as Prime and Choice. Then they organized the Better Beef Association to monitor and promote the new service.

The Secretary of Agriculture agreed to supervise the service, which would be offered on a trial basis in 10 cities for one year, provided a satisfactory agreement for its conduct could be reached by producers and packers. At an ensuing meeting, the packers claimed it was unworkable but under persuasion agreed to go along with the trial, if the National Live Stock and Meat Board would coordinate it.

Later, when the number of carcasses being voluntarily graded dropped off, it was learned that larger packers were sabotaging the program. Some were charging retailers 1 cent to 3 cents per pound more for federally graded beef or telling retailers that it wasn't available. Others had decided to offer their own house grades. Swift was offering "Premium" and "Select" to represent the two top grades. Armour followed with "Star" and "Quality."

But smaller independent packers liked the idea – federal grading permitted them to compete with the big packers. So for a few years, the Meat Board paid the salaries of four inspectors – $2 per hour – to keep the program going. By 1932, five years after the service began, there were 27 qualified meat graders on the job (including 17 part-timers) and by 1992 the number had grown to 260.

Through the years, there have been a number of changes, the major ones occurring in 1939, 1941, 1949, 1950, 1956, 1962, 1965, 1973, 1975, 1987 and 1989. In 1987, the name Good was changed to Select, to better fit consumer attitudes and perceptions as identified by the National Consumer Retail Beef Study. And in 1989, yield grades and quality grades were uncoupled, which allowed carcasses with excess fat removed on the floor to continue to be graded for quality.

Meat Import Laws Finally Work

For most of this century, when cattle prices sank, cattlemen have targeted beef imports, usually calling for higher tariffs. For decades, there was a nominal tariff.

Typically, imports fluctuated from 1 percent to 5 percent of total U.S. production. But in the early 1960s, it jumped to 10 percent, and cattlemen demanded that Congress act. (Earlier attempts at voluntary agreements had not worked.)

This resulted in the Meat Import Law of 1964. It provided for a flexible base quota based on the average production of the past three years, with a "quota trigger point" of 110 percent of the base quota. Thus, as domestic production increased, the quota also increased. On three occasions (1973, 1978 and 1979), when domestic meat prices increased substantially, the President "in the public interest" authorized imports above the trigger point, as was permitted by the law.

Photo by Charles E. Ball

William D. Ruckelshaus (left), first administrator of the EPA, visited TCFA in 1972 to study feedyard operations. Shown with Bruce Lusk (right), manager of Randall County Feedyard.

Cattlemen were infuriated, so they started a move for a counter-cyclical meat import law. After a two-year intensive effort by NCA, TCFA and others, the Meat Import Act of 1978 passed Congress on October 12 – only to be vetoed on November 12 by President Jimmy Carter, at the urging of consumer activists.

Cattlemen were really angered this time and, through incessant lobbying, gained enough support to push it through again in 1979 when President Carter signed it.

The Meat Import Act of 1979 set a floor or minimum level on imported beef of 1.25 billion pounds, which pleased importers and gave consumers protection. Through a formula, it reduces imports when domestic production increases and vice versa, which also gives cattlemen protection. Since 1980, beef imports have ranged from 7 percent to 9 percent of U.S. consumption.

Other Regulatory Agencies

Earlier in this Century, a number of other regulatory agencies were initiated that continue to impact cattle feeders, including:

- **Food and Drug Administration (FDA)**, established in 1906 to assure the public of wholesome, sanitary, unadulterated and truthfully labeled foods.
- **Federal Trade Commission (FTC)**, established in 1914 for the purpose of curbing unfair trade practices.
- **Animal and Plant Health Inspection Service (APHIS)**, a successor to the Bureau of Animal Industry established in 1884, with the authority and responsibility to control and eradicate animal diseases.
- **Commodity Futures Trading Commission (CFTC)**, established in 1974 with the responsibility of preventing price manipulation and unfair trading practices in the trading of commodity futures.
- **Environmental Protection Agency (EPA)**, created in 1970 to serve as the public's advocate for a livable environment – by integration of research, monitoring, standard setting and enforcement activities.
- **Occupational Safety and Health Administration (OSHA)**, established in 1970 with the responsibility of developing safety standards, issuing regulations, conducting investigations and issuing citations for noncompliance with safety standards and regulations.

In 1976, OSHA issued a booklet, "Safety With Beef Cattle," which became laughing stock for many cattlemen, because of its simplicity. Some sample quotes: "When floors are wet and slippery with manure, you could have a bad fall." . . . "Bare feet or sneakers aren't safe around cattle. Wear heavy shoes." . . . "Watch your step. Many people trip over animals and fall."

TCFA Response to Regulations

By 1974, the number of regulatory agencies was growing and the myriad of regulations was more than the average feedyard manager could keep up with. So TCFA hired a Ph.D., Richard McDonald, to help interpret the regulations for members, to serve as liaison with the agencies and, when appropriate, to help the agencies in writing reasonable regulations. It proved to be a beneficial move for all concerned.

In 1976, McDonald wrote that there are 15 agencies – federal and state – attempting to "watch over" the feedyard and its customers: Animal & Plant Health Inspection Service, Comptroller of Public Accounts, Environmental Protection Agency, Food and Drug Administration, Internal Revenue Service, Occupational Safety and Health Administration, Packers and Stockyards Administration, Securities and Exchange Commission, Wage and Hour Commission, Texas Air Control Board, Texas Animal Health Commission, Texas Feed & Fertilizer Control Service, Texas Railroad Commission, Texas Water Quality Board, Texas Water Rights Commission.

Oklahoma Farmer-Stockman

Cattlemen learned to burn pear for feed. But fierce droughts, like in early 1930s and 1950s, caused some to turn to the government for help.

Agricultural Adjustment Act

When Franklin D. Roosevelt was elected president in 1932, cattlemen, like everyone else, were feeling the effects of the Great Depression. Cattle prices had fallen 53 percent in three years – from an average of $10.59 per cwt. to $4.94 per cwt. In 1933, the number of unemployed came to 12,830,000 persons or 24.9 percent of the work force. Average annual income dropped to $376, down 46 percent from 1929.

As soon as Roosevelt entered office in 1933, he started implementing his "New Deal" programs, which included (1) relief for those in need, (2) recovery for business and agriculture, and (3) reform for certain economic and political practices.

A major part of the New Deal was the Agricultural Adjustment Act (AAA). To carry it out, Roosevelt appointed as Secretary of Agriculture Henry A. Wallace of Iowa, who had been editor of the family magazine, ***Wallace's Farmer***, and founder of Pioneer Hybrid Corn Company. Also, he was the son of a former Secretary of Agriculture, Henry C. Wallace, 1921-1924. After serving eight years as Secretary of Agriculture – longer than anyone – he became Vice President of the United States, 1941-1945, and a presidential nominee of the Progressive Party in 1948.

People were seeking relief, so Wallace and Roosevelt initiated many programs and agencies, most of which have prevailed to this day. A few of them include the Commodity Credit Corporation, Farm Credit Administration, Farmers Home Administration, Soil Conservation Service, Rural Electric Administration and about 20 others that do not impact cattle feeders directly.

The AAA was the first attempt at supply management. It paid farmers to (a) reduce acreage, (b) store commodities on the farm, and (c) established price supports based on a new concept, parity. Parity was a method of measuring agricultural equality (based on 1910-1914 prices) and a program of price supports to achieve it.

The AAA program was financed by a processing tax. But a decision by the U.S. Supreme Court in 1936 largely ended the AAA, denying federal power to regulate production and levy processing taxes for this purpose. Roosevelt prevailed on Congress to pass a new Agricultural Act of 1937 and another in 1938, however, and continued some of the programs that exist today.

Early in 1933, as Congress debated the AAA, cattlemen opposed all efforts to have cattle included as a basic commodity, which would have made them eligible for production controls and price supports. They stuck by their basic philosophy of independence and free enterprise.

But by 1934, the Plains states were experiencing the fiercest drought in all history – in addition to the worst depression. Dust storms were so bad that some towns burned

Beginning June 1, 1934, the government purchased 8.3 million cattle for an average of $13.50 per head – and shot most – reducing the U.S. cattle population 11 percent in eight months.

lights by day. Grasshoppers and rodents devoured feed supplies. Water holes gave up and wells quit. Many cattlemen hauled water – by wagon, trailer or truck – but they couldn't keep ahead of the thin and thirsty cattle. Cattle were dying by the thousands and cattlemen began to reconsider their philosophical stance.

As early as 1933, Dolph Briscoe, president of the Texas and Southwestern Cattle Raisers Association (TSCRA) and father of the future governor of Texas (Dolph Briscoe, Jr., 1973-1979), urged that an orderly marketing program be instituted.

From 1928 to 1934, cattle numbers increased 20 percent – to 68 million head, the largest number in history – and few consumers had the money to buy beef.

On December 2, 1933, the Northeast Panhandle Feeders Association called a meeting of about 500 cattlemen in Canadian to discuss what might be done about "the present deplorable situation." A committee – comprised of H. E. Hoover, Canadian; W. D. Price, Pampa; B. F. Jackson, Miami; Grover B. Hill, Amarillo; and Gene Howe, Amarillo newspaper publisher – was appointed to draft resolutions expressing the views of those present. Among their recommendations to the government: (1) reallocate appropriations so that beef and dairy industries will receive equal benefit; (2) cattlemen should receive as much consideration from the government as producers of wheat, cotton and other commodities; and (3) cattle should be classed as a basic commodity, thus qualifying for government support.

The next month, January 1934, the American National Live Stock Association met in Albuquerque, N.M., where the dominant topic was whether beef should be made a basic commodity. It was so controversial that it never came to a vote. But immediately following the ANLSA meeting, Dolph Briscoe, president of TSCRA, called a meeting of Texas cattlemen present. Jay Taylor of Amarillo made a motion that Texas support the bill then before Congress – the Jones-Connally bill – amending the Agricultural Adjustment Act to include cattle as a basic commodity. It passed 40 to five. So a telegram to that effect was sent to Texas senators and congressmen. The Jones-Connally bill – written by two Texans, Rep. Marvin Jones of Amarillo and Senator Tom Connally of Marlin – appropriated $63 million for cattle purchases by the Federal Surplus Relief Corporation. (The total aid figure later was boosted to $525 million.)

So on June 1, 1934, the Emergency Cattle Purchase Program began. Part of the cattle were slaughtered (the meat canned by relief workers and given to the poor) and many were shot on the spot.

During the following eight months, the government purchased 8.3 million head, reducing the U.S. cattle population 11 percent. The average price paid was $13.50 per head. Of the total purchased, 1,908,570 were in Texas; 487,154 were in Oklahoma; and 544,922 were in New Mexico.

It remained a controversial program. Evetts Haley, rancher and historian from Canyon, wrote in an article, "Cattle Business and Monkey Business," that participating ranchers "surrendered their birthright for a herd of canner cows."

But the ***Canyon News*** responded in an editorial, "Most Texas ranchers wanted aid of some kind, not rugged individualism."

At its 1935 convention in Houston, the Texas and Southwestern Cattle Raisers Association passed a resolution of appreciation for the government aid: "The Government cattle buying program furnished a market for thousands of cattle which would otherwise have perished. This program . . . furnished work for thousands of unemployed; and . . . was the means of removing surplus supplies (of livestock) and thereby contributing to restoring the cattle industry to a more stable basis."

In any case, by 1935, drought, dust and government purchases had ended the problem of overproduction.

THE DAY THEY SHOT OUR CATTLE

As a 10-year old boy on a small East Texas farm, it didn't take a lot to impress me. But one thing that really impressed me was the day they shot our cattle.

We had about 30 head, counting calves, which we fed and named and occasionally rode in a kids rodeo. In the summer of 1934, it was hot and dry; the sky was occasionally clouded with dust that blew in from West Texas; our pastures were extremely short; the cattle were unseasonably thin; the creek had dried up; and our three ponds were mere mud holes. We were hauling water in barrels in a wagon – water which we drew by hand from a neighbor's well – but the cattle just stood around the water trough, always thirsty.

Obviously, they should be sold. Everybody wanted to sell. But there absolutely was no market. And many cattle were dying from starvation.

President Roosevelt's "New Deal" programs, through the AAA, already were offering relief to farmers who would plow up cotton (we did) or kill little pigs (we didn't), but independent, strong-willed cattle industry leaders were holding out. They had too much pride and self-determination to accept government help. But when you run out of water, pride and self-determination tend to wither, also.

Finally, in June 1934, we heard on the radio that Congress had approved cattle as a basic commodity and appropriated $63 million for a purchase and slaughter program. Cattle in good shape would be purchased, slaughtered in local abattoirs, canned by people on work-relief and given to the poor. Cattle that were too thin (like ours) would be condemned and shot on the spot.

For several weeks, my dad and mom agonized over whether to have our cattle shot. Dad pointed out that the government was offering (for condemned cattle) $12 per head for those over 2 years of age, $10 for yearlings, and $4 for calves, "pretty good, considering these times." Furthermore, the price included a "benefit program" for producers with mortgaged cattle, which meant the banker wouldn't get more than half the payment. For cattle not condemned, the price was much better: $12 to $20 for those over 2 years, $10 to $15 for yearlings and $4 to $8 for calves.

As we waited, the cattle were getting poorer, we were pulling more out of mud holes and hauling water was becoming "less noble." So my dad gave in.

Two government agents arrived late in the afternoon with high-powered rifles; we had all our cattle penned in a lot right behind our house; and the big man simply said, "Which ones?" Dad would point to one – starting with the weakest and ugliest – and they would shoot. After about 12 deads, it became harder and harder for my dad to point. In less than one hour, the men left, leaving 18 dead cattle in our lot. Neighbors came and took two of the fatter calves to dress that night. The next day, we drug the remaining deads to the back of the pasture, where the buzzards and coyotes had a feast.

A lot of other cattlemen went through similar experiences. Within eight months, the government purchased 8.3 million head, reducing the U.S. cattle population by 11 percent. The average price paid was $13.50 per head.

– Charles E. Ball

Photo by Charles E. Ball

Paul Marion was superintendent and beef researcher from 1941-73 at the Spur (Texas) Experiment Station, established in 1906 to "find ways to utilize West Texas farm products." Ca. 1962.

Chapter 5

BUILT ON TECHNOLOGY

For 100 years, research has paved the way for cattle feeders.

Nature designed cattle for the open range, and getting them to perform well in the feedyard has always been a technological challenge.

For more than 100 years, that challenge has evolved around nutrition, animal health, genetics, meat science, mechanization and transportation. As progress was made in those fields – and progress is still continuing – technological endeavor moved into new fields, such as risk management, computers, communication and the environment.

New USDA, Land Grant Colleges

Agricultural technology got its greatest boost in 1862 when Congress passed two landmark pieces of legislation. One established the U.S. Department of Agriculture. The other provided for land grants to each state – 30,000 acres for each senator and representative it had in Congress – which was the origin of land-grant universities. Altogether, the state and territories received 11,367,832 acres, which could be sold, the proceeds invested and the income used to create a college for agriculture and the mechanical arts.

In the years that followed, these institutions would serve as the centers for research that would provide most of the science now responsible for modern cattle feeding.

Texas A&M College was established in 1876, followed by the Texas Agricultural Experiment Station (TAES) in 1888. One of the initial TAES projects, in 1888, dealt with the feeding of cattle in drylots. Since then, hundreds of projects have been conducted that relate to cattle feeding.

First Research in Oklahoma

At Oklahoma A&M College, the first cattle feeding project was in 1894 by Alexander C. Magruder. He fed 12 head of Shorthorn steers for 110 days to a weight of 1,050 pounds, and sold them for $4.10 per cwt. in Kansas City. Ironically, this cost him his job. Magruder was "a Mississippi-born Democrat" and he got caught in some Oklahoma political crossfire, according to Donald E. Green, in ***A History of the Oklahoma State University Division of Agriculture***. It seems that in 1893 President Grover Cleveland had replaced the Republican governor of Indian Territory with a

Photos by Charles E. Ball

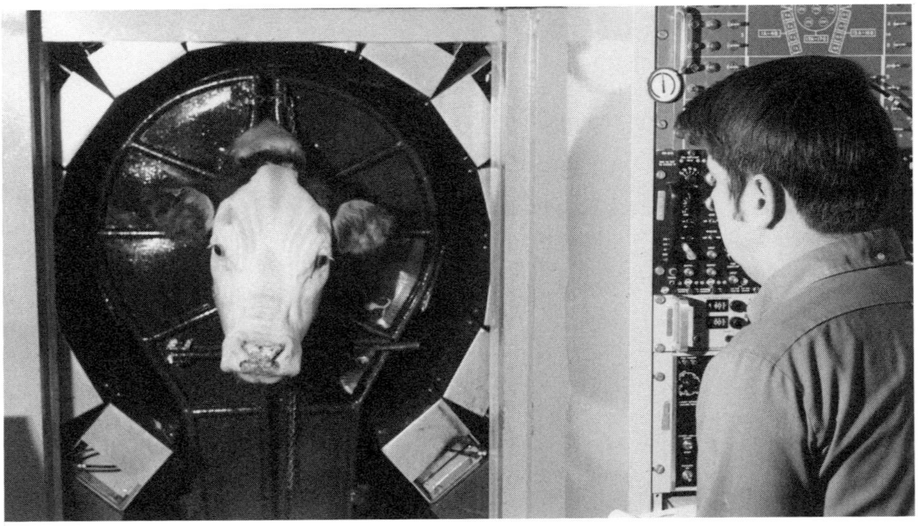

In the 1960s, Dr. Lowell Walters (left), OSU meats researcher, was advocating less fat on beef animals, "because (1) it's a waste of feed and (2) consumers are beginning to resist it." Ahead of its time, the K-40 counter at OSU (right) was measuring lean-to-fat ratios in the 1960s. Basis: Most of the potassium in the body is concentrated in the muscle tissue.

Democrat, which upset the Board of Regents of the new Stillwater college (established in 1890), and the new college administration became embroiled in the controversy. Magruder told the Board that he had sold the cattle for $4 per cwt. to a local butcher, which they approved, but then the butcher reneged. So he took the cattle to Kansas City, where he visited the packing plant to get the packer's opinion of the quality of the dressed beef.

When Magruder submitted a travel claim, the Board considered the trip to Kansas City a violation of its wishes. Magruder's explanation that the original buyer reneged, that he got $.10 per cwt. more than the original bid and that he also obtained "carcass data" from the packer was insufficient for the Board, so they fired him.

Since Marguder was the only professor of agriculture, except the director of the Experiment Station, this "scandal" set back research at Oklahoma A&M for some time.

The first significant research of cattle feeding in Oklahoma was by Warren L. Blizzard, during the 1920s. In 1926, he founded the annual Feeders Day, which became a major social event, as well as an educational program, until it was discontinued in 1970. A unique feature of the popular Feeders Day was the "Free Feed," prepared and served by Clay Potts. Initially, he served strip steaks and later pit barbecue. He had to modify the entree, because in the late 1940s and early 50s he would serve as many as 3,000 to 4,000 people – 10 times the number attending the educational programs. Students, townsfolk and their families would show up – for lunch only!

Early Research in New Mexico

The first cattle feeding research in the Territory of New Mexico, which became a state in 1912, was in 1895. At least, it was in 1895 that Bulletin 17, entitled "Principles of Stock Feeding and Some New Mexico Feeding Stuffs," was published. The Agricultural Experiment Station of Las Cruces College – later to become New Mexico A&M College and still later New Mexico State University – was established in 1889. But there was no agriculture department, so the feeding research was conducted by the chemistry department.

Photo by Ralph Leone

***Feeds and Feeding**, which became a bible for animal nutrition students, was first edited by W. A. Henry in 1898, with help from graduate student F. B. Morrison.*

First Books on Nutrition

The roots of animal and feeding science were set more than a century earlier, however, with the discovery of nitrogen in 1772 and of amino acids in 1810.

Justus van Liebig of the University of Giessen, Germany, published his book, ***Organic Chemistry in Its Application to Agriculture and Physiology***, in 1840, touching off a wave of interest in organic chemistry that dominated agricultural science for the rest of the nineteenth century.

By 1880, Henry P. Armsby, a chemist at the University of Connecticut Agricultural Experiment Station, published what may have been the first system of balancing rations in the U.S. But there was still so much to learn. For example, Armsby warned feeders against allowing their cattle access to too much water; and he believed that too much oxygen would inhibit fattening (a belief that led to the bleeding of animals in Germany). He also cautioned that "most feed matters have not been tested, only estimated."

Before the turn of the century, W. A. Henry, dean of the School of Agriculture at the University of Wisconsin, edited the first edition of ***Feeds and Feeding***, in 1898.

A clue to Henry's wisdom, which would appeal to students and cattlemen alike, was in the book's preface: "The stockman whose training has been solely in the school of experience often holds in light regard that which is written concerning his vocation. Let him remember that facts and truths are the same whether their repository is a book or the human mind. Held by the latter, all perish with the possessor; in the keeping of the former, the whole world may be benefitted.

"On the other hand," he continued, "novices usually underestimate the importance of experience, often thinking that by reading they can acquire the knowledge necessary in the persecution of their business. That stock feeding is an art and not a science, and that experience and judgment must rule in its successful conduct, is recognized in this volume by placing on the title page the adage, 'The eye of the master fattens his cattle.'"

In a chapter, "Counsel in the Feed Lot," Henry said: "Indian corn must continue the great grain food for steer fattening in the United States. . . . The success of steer feeding in America must depend largely upon the supply of Indian corn available for this purpose."

Later, he added, "Corn is never so acceptable to a steer as when unhusked. There is a freshness and palatability about an ear of corn wrapped in Nature's covering which every steer recognizes and shows by the eagerness with which he consumes it."

Henry was succeeded as editor-publisher of ***Feeds and Feeding*** by Professor Frank B. Morrison in 1915. The book, updated frequently until 1956, became known as the "Nutrition Bible" among the countless students who studied it as a college text book and used it as a reference book after leaving college. Over a million copies were sold through the years and it became one of the most widely-used reference books in the world.

Major Discoveries

Along the way, some major discoveries were made that impact cattle feeders yet:

■ Corn shellers and hammer mills were invented about 1840 and began to improve feeding efficiency. The new technology of grinding – often by mobil "custom" grinders with regular customers among farmer feeders – was spurred further by high grain prices.

■ The refrigerator car was invented in 1868 and mechanical refrigeration was well established by the 1890s. This revolutionized the meat packing business. It could now be a year-around business, as could cattle feeding.

Photo by Monk Vance

A leader in meats research, 1947-78, Dr. O. D. Butler of Texas A&M, pioneered research in cutability, tenderness, palatibility, and electro-stimulation.

■ In the 1930s, before the advent of home refrigerators, refrigerator locker plants were labeled "revolutionary." The USDA defined them as "a safety deposit box, kept in a room with a temperature near zero Fahrenheit, where families may freeze and store food supplies which have been grown at home or bought." An ***Advertising Age*** survey in 1938 revealed that 2,500 locker plants had sprung up in 21 states, offering 850,000 lockers, each capable of storing 250-300 pounds of meat.

■ Breeding improvement began in Texas in the 1870s, when ranchers Captain Mifflin Kenedy and Col. Charles Goodnight first brought in Shorthorn bulls to upgrade the native Longhorns. Then crossbreeding, begun by the King Ranch in 1924, showed improved feedlot performance for up to 120 days from Shorthorn-Brahman and Hereford-Brahman steers. Later, the exotic breeds started arriving – first Charolais in the 1950s and Simmental in the 1960s – until today the U.S. has over 30 registered breeds, not counting the crosses.

■ Early in the 20th Century, scientists and farmers discovered that younger animals gained more efficiently than the three- and four-year-old steers traditionally fed out. This and the high priced grain led to the introduction of "Baby Beef." (See page 24.)

■ Silage was introduced in the 1870s and 1880s. Early on, corn harvested as silage was chopped manually and packed by teams of horses, mules or oxen.

■ "Texas fever" not only was causing huge economic losses in the late 1800s but caused 15 states to close their borders to Texas cattle. (From 1866 to 1895, some 10 million Longhorn cattle were driven to out-of-state markets.) In 1892, Texas A&M researchers concluded that the fever was carried by ticks; in 1889, Robert J. Kleberg of the King Ranch built the first dipping vat; in 1906, a quarantine line was drawn across the state and the vast dipping program began. Over 5,000,000 cattle below the line were dipped the first year, and the number increased until 1921 when 15,367,208 were dipped. By 1938, the state was declared "clean."

■ Vitamin A and B were discovered in 1915 and recognized as important to animal health.

Historical and Biographical Record of the Cattle Industry

First dipping vat was built in 1889 by Robert J. Kleberg of the King Ranch for control of "Texas Fever."

Fort Worth Star-Telegram Collection, University of Texas at Arlington Library

Early research at the Spur (Texas) Experiment Station was conducted on grain sorghum, cottonseed products and vitamin A. Among the scientists there, between 1916 and 1940, were J.M. Jones (left), R.E. Dixon (center), and J.K. Riggs (right).

■ Trace minerals were made known in the 1920s; and during the 1930s feeders began using additives in their feeds.

■ Urea became important in some rations after World War I.

■ Vitamin B12 was discovered in 1948 and led to the discovery two years later that certain antibiotics added to feeds could increase growth.

■ Diethylstilbestrol (DES), the first important growth hormone, was isolated in 1950 and approved for use in food animals in 1954. Though banned in 1979 as a possible carcinogen, DES increased weight gains up to 19 percent and quickly introduced the feeding industry to the concept of using hormones in animal production.

Early Research in Texas

The USDA started investigations on feed grains at Channing, Texas as early as 1903. This was in cooperation with management of the Capitol Freehold Land & Investment Co., who wanted better grains for the cattle on their XIT ranch. The experiment station remained at Channing until 1905, when it was moved to Amarillo. Citizens of Amarillo put up $5,000 to have the station relocated in their town (about where the Tri-State Fair would later be located). It was abandoned in 1920.

By 1913, USDA officials were seeing some possibilities in the Southern Plains for "feeding cattle at home rather than sending them out of state," according to ***The Cattleman*** magazine. So they sent "an agent to what is commonly known as the Panhandle of Texas" to help farmers and cattlemen learn to "feed cattle for market." His first demonstration, on a farm near Amarillo, showed that steers fed cottonseed meal, kaffir corn silage and sorghum "made a net profit of $7.39 per animal."

"Following this demonstration, much time was spent in encouraging ranchmen to put in pit silos and in instructing them in the growing of silage crops," continued the magazine. "During the winter of 1914-15, about 7,000 head were fed in accordance with instructions furnished by the U.S. Department of Agriculture."

Soon afterwards, in 1918, R. E. Dixon, superintendent of the Experiment Station at Spur, Texas, conducted the first research on feeding grain sorghum. "At that time, he foresaw the possibilities of a cattle feeding industry in the Rolling and High Plains of Texas," said Paul T. Marion, superintendent of the Spur station, in a speech to the first Cattle Feeding Conference of the TSCRA Cattle Feeders Division in 1961. "He worked with a number of leading animal husbandmen (at the station), including J. M. Jones, John H. Jones, Dr. J. L. Lush and John K. Riggs in an effort to provide information on the utilization of feeds produced in the area."

Did cottonseed meal "poison" cattle? After many tests, researchers at Spur, Texas concluded it did not. This steer was fed cottonseed meal and hulls, white grain and 1 pound of alfalfa per head per day for 550 days. It sold on Sept. 14, 1936, weighing 1,350 pounds, for $8.50 per cwt.

Texas Agricultural Experiment Station

Solution to Night Blindness

It was at Spur that John K. Riggs (a retired professor from Texas A&M) participated in the historic research on vitamin A and "night blindness" or "meal poisoning" or "fat sickness" in feedlot cattle. This was a condition that many cattle developed when confined and fed cottonseed meal. It was a tricky problem, because not all cattle got the malady and researchers could not duplicate it with any consistency.

It was not a new problem. In the Bible, Jeremiah 14:6, reporting on "the great drought" 600 B.C., states: "The wild asses stand on the bare heights, they pant for air like jackals; their eyes fail, because there is no herbage." This indicated early knowledge of the effect of drought or lack of green feed upon the well-being of animals.

In 1935, researchers finally made the link, according to Riggs. The problem centered on the availability of vitamin A in the feed. Cattle from different backgrounds had different amounts of the vitamin stored in their bodies, which accounted for the erratic problem. The solution was simply to feed minor amounts of products rich in carotene or vitamin A.

Southwest Collection

Dr. W.L. Stangel, head of animal husbandry at Texas Tech, started cattle feeding tests in 1925 – mostly with cottonseed products.

Research at Texas Tech

After Texas Tech was founded by the Texas Legislature in 1923, Dr. W. L. Stangel became the first head of Animal Husbandry (1925) and began advocating cattle feeding as an outlet for West Texas crops. As soon as Stangel arrived at Tech, a local farmer, H. W. Stanton, asked him to formulate a feed ration, using products off his farm – grain sorghum heads and cottonseed meal. With this ration, Stanton started a feed business (Standard Milling Company) and in 1931 a feedyard on the rim of Yellow House Canyon for 3,000 steers.

By 1933, Stangel was conducting experiments to compare cottonseed meal with whole cottonseed. For the next three or four decades, feeding research was sponsored by the National Cottonseed Products Association and the Texas Cottonseed Crushers Association. (Tech never received research funding from the state or federal government; instead, the researchers had to depend on private contributions or profits from their research.) Stangel later became Dean of

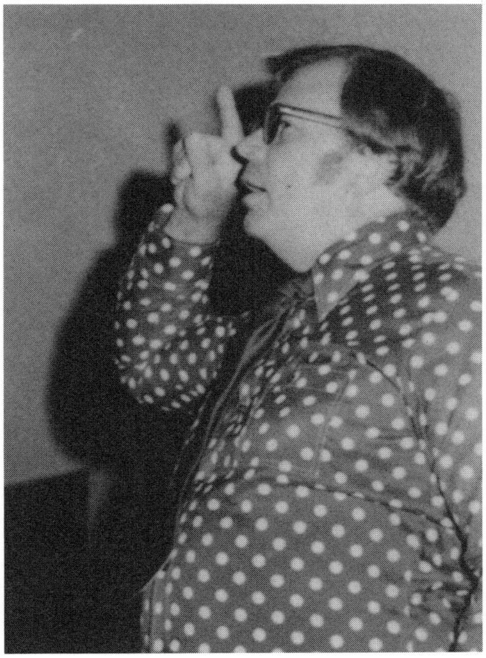

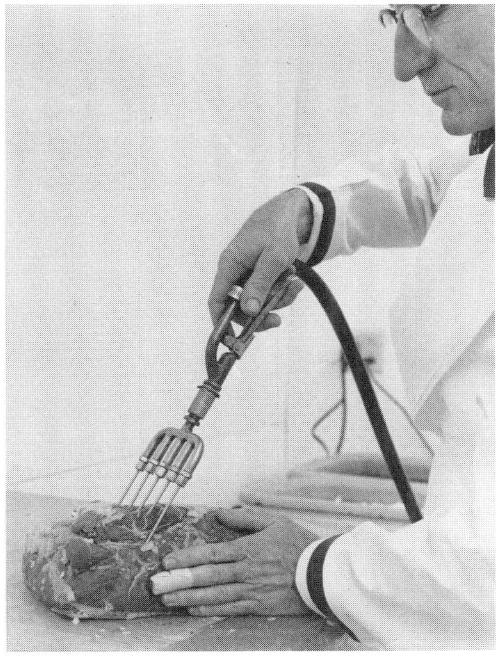

Photo by Charles E. Ball

Innovative Dr. Ralph M. (Bull) Durham developed a machine to add artificial marbling to beef and conducted other controversial research at Texas Tech, 1959-80.

Agriculture at Tech (1945-58), and never stopped promoting cattle feeding in West Texas until his death in 1978.

Following World War II, Tech acquired the Pantex Ordinance Plant north of Amarillo and renamed it PanTex Farm, later Texas Tech University Research Farm. J. P. Smith was the first superintendent (1947-63), conducting research on wheat pasture poisoning of cattle. With a $530,000 grant from the Killgore Foundation (founded by C. L. Killgore, an early rancher and feeder in the area), the Killgore Beef

At the Killgore Beef Cattle Research Center, east of Amarillo, TCFA sponsored on-going research, 1968-80. Dr. Lloyd Sherrod, superintendent (right) and Dr. Richard McDonald, TCFA.

Cattle Center was added in 1964. And it was here that early TCFA-sponsored research was conducted, during the late 1960s and 1970s, much on shipping fever.

Another creative, albeit controversial, researcher at Texas Tech was Dr. Ralph M. (Bull) Durham, who was head of the Department of Animal Husbandry, 1959-65, and on the staff until 1980. He persistently pushed for "profitability factors" in commercial cattle and bluntly criticized "show ring characteristics," which kept him in hot water with purebred cattlemen. Among the Durham innovations was crossbreeding of dairy bulls on purebred Angus cows, resulting in "Holgus" bulls for range use; artificial marbling, created by the injection with a multi-pronged needle of either animal fat or vegetable fat into lean meat; and the feeding of manure back to cattle.

Another innovation, which probably was Durham's greatest contribution to cattle feeding, was all-concentrate rations. A number of feedyards tried all-concentrate and liked it, especially for light cattle. Most settled for "high concentrate," however, and gave Durham credit for the trend.

Texas Agricultural Experiment Station

Hybrid grain sorghum gave cattle feeding in the Southwest its biggest boost. Roy Quinby (left) and J.C. Stephens of the Chillicothe Experiment Station generally are credited with the development, in the 1950s.

Historic Hybrid Grain Sorghum

Another piece of historic research – possibly the single most important in the evolution of cattle feeding in the Southwest – was the introduction of hybrid grain sorghum. Sorghum originated in Africa and proved to be far more adapted to dryland production on the semi-arid Great Plains than was the native American grain plant, corn.

The first cultivated sorghum reached Texas in 1857, and research on sorghum began in 1889. The original Milo Maize, also known as Giant Milo, was so tall it was harvested by workers – cutting off each head with a pocket knife – while standing in a wagon pulled through the field by mules. By World War II, dwarf and double-dwarf varieties were developed, suitable for harvest by combine.

The USDA and the Texas Agricultural Experiment Station (TAES) began research on grain sorghum at Chillicothe in 1905. While several scientists through the years researched the concept of hybrid grain sorghum – a tedious task because both the pistil and stamen are in the same spikelet – the two who generally are credited with the development were Roy Quinby and J.C. Stephens of the Chillicothe station.

In 1956, the TAES sold parental seed to seed growers – enough to plant 10,000 acres – which triggered the boom in grain sorghum production. Yields jumped up to 40 percent. Irrigation, just becoming widespread on the South Plains, further boosted yields – up to 7,000 and 8,000 pounds per acre. For the state as a whole, including dryland and irrigated, average yields increased from 1,064 pounds in 1951 (a pre-hybrid year), to 2,520 pounds in 1961, to 3,472 pounds in 1981.

(Continued on page 81)

Wyatt McSpadden

Wyatt McSpadden

Bill Rhew

Wyatt McSpadden

David Langford

Wyatt McSpadden

David Langford

Wyatt McSpadden

Wyatt McSpadden

Wyatt McSpadden

Wyatt McSpadden

Wyatt McSpadden

David Langford

Courtesy Continental Grain Co.

Courtesy Caprock Industries

Holly Kuper

Wyatt McSpadden

Courtesy CDH Feeders, Inc.

Courtesy CDH Feeders, Inc.

David Langford

David Langford

Wyatt McSpadden

Wyatt McSpadden

Ralph Leone

Burt Rutherford

Photo by Charlrs E. Ball

Chardo W. Pierce, Lubbock, invented and promoted a micronizer – utilizing infrared gas burners – in the 1960s, which increased feed efficiency of grain sorghum 10-15 percent. Most feedyards preferred steam flaking, however, which produced similar results.

Mountains of grain sorghum resulted, and nowhere to market it. Now was the time for longtime advocates of cattle feeding in the Southwest to make their move. And they did, along with the new Grain Sorghum Producers Association, organized in 1955 and headquartered in Amarillo.

Early tests showed the feeding value of grain sorghum to be about 95 percent that of corn. But that was before the new methods of processing grain became available. Steam flaking, introduced to California feedyards after World War II, increased feed efficiency 10-20 percent, but was found to be an expensive process. Steam flaking equipment was costly, had high power requirements and was relatively slow. So during the 1960s and 70s, feedyard operators tried several alternatives – pressure cooking, reconstituting, popping, micronizing – all of which gave comparable results, 10-20 percent increase in efficiency. Nevertheless, by 1990, most commercial feedyards settled on steam flaking.

Eradication of Screwworms

Still another technological development of historic proportions – and possibly the most dramatic ever – was the eradication of screwworms, the most repugnant pest to ever attack livestock. With any break in the skin, the screwworm fly would lay eggs onto the wound or blood, the eggs would hatch and the larvae would bore into the flesh until the animal was treated or died. Cattle losses in the southern United States were exceeding $15 million a year.

The technology was dramatic, because it involved atomic energy and the sex life of the screwworm fly. Charlie Ball in 1961 was asked to write a story, "Atomic War on Insects," for the ***Saturday Evening Post*** and it was reprinted in the ***Reader's Digest.***

Capsuled, here's the story: In 1938, two young scientists at a USDA laboratory in Menard – Edward F. Knipling and Raymond C. Bushland – observed that the

Photo by Charles E. Ball

Dr. David P. Hutcheson, nutrition researcher with Texas Agricultural Experiment Station in Amarillo, 1992, shows pinpointers (at Bushland), partially financed by TCFA. The pinpointer records on a central computer the time and amount of feed consumed by each animal.

screwworm fly mated only once in its lifetime. "If we could just sterilize the males," commented Knipling to Bushland, "we could trick them into breeding themselves out of existence." The technology – atomic irradiation, which permitted the sterilization – was not available until after World War II and a long struggle for funding. Through the Southwest Animal Health Research Foundation, cattlemen raised $3 million – with contributions of 50 cents per animal unit – to launch the program. Fly factories were built to rear, irradiate and release 50,000,000 flies per week. And in 1966, the United States was declared "screwworm free."

Commercial Technology

Not all the new technology has come from public institutions. Feedyards themselves in many cases have served as large laboratories for the testing and proving of new ideas. Most feedyards have retained consulting veterinarians and consulting nutritionists (usually with Ph.D. degrees) to guide, assist and train their employees.

And in recent years, as government funds became more limited, commercial companies have stepped in with some of the most advanced technology, especially in the animal health area. Pharmaceutical companies report that it now takes seven to 15 years to get a drug developed and marketed, at an average cost of $3 million to $30 million. Animal health has grown into a $3 billion market.

As cattle feeding evolves and progresses, there is no doubt that technology still holds the key to the future of the industry. That's the way it's always been. That's how it will continue to be.

Library of Congress

Introduction of the two-wheel trailer, followed by pickups and cattle trucks, permitted farmers and ranchers to haul (rather than drive) their cattle to market . . . and ushered in the era of livestock auction markets. San Angelo Stockyards Auction, 1939.

INTRODUCTION OF THE TWO-WHEEL TRAILER

When A.L. Black was growing up in New Mexico during the 1920s and 30s, he occasionally had the privilege of riding the train to Kansas City with his dad's fed cattle. He later moved to Friona, Texas, where he became a successful farmer, agricultural leader, founder of Friona Feedyard, chairman of Friona Industries, and a director of TCFA. Now, he tells how shipping cattle by train and subsequent transportation developments have impacted the cattle industry:

"The mode of transportation has always dictated the way we market cattle. First there were cattle drives–the only way to deliver cattle. Then the railroads arrived (around 1900 in this region) and we drove cattle several miles to the railheads, for the beginning of their trip to the central markets – Fort Worth, Kansas City or Chicago, in our case.

"In 1921, my dad bought a new Dodge, which offered a new challenge for transporting cattle. I remember the first two-wheel trailer in our country. We took the front axle and rubber-tired wheels from an old buggy, built a wooden bed and hooked it to the Dodge. A few days later, we loaded our old milk cow, which was for sale, and started the 30-mile trip to Clovis. But the trailer broke down – the 'high speed' jolting was too much for the old buggy wheels – and we ended up driving the cow the last 15 miles.

"Later, about 1937, we traded for an old pickup. But it wasn't big enough to haul cattle. So we took the bed off, added a fifth-wheel, built a 14-foot trailer, and that was the first semi-trailer truck in our area. It was the biggest motor vehicle I had ever seen!

"Shortly after World War II, a cattleman near Bryan had a novel idea – a goose-neck trailer, to be pulled by a pickup. And 30 years later, nearly every cattleman had one.

"At that same time (late 1940s) the semi-trailers were single-axle, 34 feet long. And by the 1970s, we began to see 18-wheelers, with double-deck pot-bellied trailers, 50 feet long, that carry up to 50 fed cattle or 120 feeder calves.

"But it was the two-wheel trailer that had the biggest effect on how we market cattle. It permitted farmers to haul our cattle to market, rather than drive them. And that made feasible the local livestock auction market. The first ones in this area appeared about 1930; and by the end of that decade, there was an auction market in nearly every county or at least within a two-hour drive for every pickup or trailer.

"By 1952, there were 2,500 livestock auction markets in the country. They, along with direct buying by packers, soon put the big central markets out of business."

George Diedrichsen (left) and son Howard moved from Nebraska to Dumas, Texas, and with a group of local investors built Moore County Feeders, Inc. Shown at opening, October 1960. Although the feedyard ownership changed, Howard was manager 1960-92.

Chapter 6

ENTREPRENEURS AND THE BOOM

"The cattle feeding boom – greatest thing since the oil boom of the 1930s."

The boom. It surfaced in 1959, but its seeds were planted in 1957, with the arrival of hybrid grain sorghum on the High Plains. In 1956, Texas produced 124 million bushels of grain sorghum. The next year, with hybrids, 244 million.

The price dropped and a lot of High Plains farmers started looking for help. The avenue they chose was the Grain Sorghum Producers Association (GSPA).

GSPA went immediately about looking for ways to sell more grain, and the board set three goals: (1) increase exports, (2) feed more cattle and (3) feed a greater percentage of sorghum to the cattle which were fed.

At the time, the common thinking was that grain rations should contain no more than one-third milo. Other grain needs were usually met with corn, which was more costly but supposedly better for the cattle.

Packer Prejudice Against Milo-Fed Cattle

A. L. Black, veteran cattle feeder at Friona, Texas, tells about the packer prejudice against milo-fed cattle that had existed for years. "We used to ship our cattle into Emporia," he says with a wry grin, "feed 'em for 48 hours on corn – so they'd have corn in their droppings when they got to Kansas City – and they'd be acceptable."

Most of the Panhandle's milo was shipped west, where it was used widely, though sparingly, in the feed bunks of the large commercial cattle feeding industry of the West.

D.G. (Bill) Nelson of Amarillo, hired by GSPA first as a consultant and later as executive vice president, recalls the Association first sponsored research through Texas Tech's Dr. Ralph Durham. The research showed that high levels of milo could indeed be fed safely to cattle, and compete quite favorably with corn. (Similar research had been done earlier by the Texas Agricultural Experiment Station at Big Springs and Spur, and by Oklahoma A&M College.)

That information laid a still-stronger groundwork for a Texas feeding industry to build on, for at the time there was virtually no irrigated corn production in the Panhandle. In 1959, GSPA and the West Texas Chamber of Commerce (WTCC) announced the first of three tours which would eventually change the face of the Panhandle.

Dean W. L. Stangel, former dean of agriculture at Texas Tech and chairman of the WTCC Agriculture Committee, led the first tour. He had been a long-time advocate of

Amarillo Globe News

Southwest Collection

Amarillo Globe News

Major promoters of cattle feeding on the South Plains during the boom years included (from left to right): Sam Thomas, manager of agricultural development, Southwestern Public Service; Dean W. L. Stangel, chairman of the Agriculture Committee, West Texas Chamber of Commerce; and D. G. (Bill) Nelson, executive vice president, Grain Sorghum Producers Association.

more feeding in the Panhandle, believing the region had the capability of developing, with milo, the same kind of farmer-feeder complex which had served Cornbelt farmers so well. In 1951, for example, he wrote: "There are 16,903 farms in these 16 South Plains counties. If each farm feeds out one carload, or 20 head of cattle yearly, the South Plains could feed out 338,000 head! This can be compared to the number of cattle now being fed in the three leading Corn Belt states: Iowa, 620,000; Nebraska, 260,000; Illinois, 222,000."

The 1959 tour visited Illinois and Iowa, and the 78 participants were not too impressed with the small, labor-intensive feedlots they saw. That first tour yielded few results.

West Texans Tour the West

Things would change with the next tour, however, which covered the commercial lots of the West. The second tour, in 1960, was billed by the WTCC and GSPA as a "Let's Visit Our Customers" tour. Nelson says the thought that West Texans might decide to build their own carbon copies of the commercial yards had occurred to the milo salesmen, "But if we'd come out and said to the bankers and farmers, 'Let's go see how to build feedyards,' we'd never have gotten them to go."

Anyway, 204 people did make the tour of Arizona and Imperial Valley of California. In 1962, on a third tour, another 183 visited the San Francisco-Bakersfield areas.

Walter Lasley of Stratford, who already had started his feedyard (in 1953) and was looking for expansion ideas, made both of the California tours. "The first time, everybody rolled out the red carpet for us," he says with a grin. "But the second time, they had become leery and only the Extension people greeted us, as a matter of courtesy."

The tours, conceived primarily by grain farmers, and businessmen like Sam Thomas, agricultural manager for Southwestern Public Service Co. of Amarillo, had been designed simply to help move more grain, to provide markets for the surpluses piled in Texas granaries.

They did that, but they did much more. The feedlot boom of the 1960s can be traced back to those tours, for the people who made them came home and with few exceptions contributed to the phenomenon of the Panhandle – bankers by lending money and pushing for new feedyards, farmers by becoming stockholders in new ventures, businessmen by developing corollary industries to help support the feedyards.

Tours by West Texas grain farmers, agricultural leaders and would-be cattle feeders to Arizona-California in 1960 and 1962 fueled the boom. The tourists gained new ideas and later attracted entrepreneurs from the Far West.

Spectacular Growth

The boom was ignited. Writers and speakers waxed poetic about the growth of the beef feeding industry in Texas during the 1960s. "Spectacular," "the miracle on the Plains," "the greatest Cinderella story in agriculture," they called the growth, particularly in the Texas Panhandle. Gene Benham, president of the First State Bank of Morton, called the cattle feeding business in his area, "The greatest thing since oil came to this country back in the 1930s."

Predictions about the future growth of the feeding industry were not uncommon, like this one by Jay L. Taylor, Amarillo cattleman-banker, in a 1965 speech to bankers at Texas A&M University: "In less than 10 years we will be feeding right here in Texas all the calves and yearlings raised in Texas. So demand for credit will be tremendous. Get ready."

With fed cattle numbers ranging from 300,000 head in 1958 to nearly 2 million in 1968 and 4.8 million in 1973 – a 1,600 percent increase in 15 years – "spectacular" had to be something of an understatement.

Charles E. Ball, the new executive vice president of the Texas Cattle Feeders Association, in a 1973 speech gave five reasons for the boom in the Texas Panhandle and in neighboring Oklahoma, Kansas and New Mexico: (1) availability of feeder cattle, (2) availability of grain, (3) relatively good year-around climate, (4) financing and (5) entrepreneuring people.

Although physical changes were made in existing Texas feedyards, few of the feeders had experience at feeding cattle year-around on a large scale (An exception was Durward W. Lewter, who by 1960 was feeding 30,000 head at Lubbock.) That background soon came from outside the area.

Out-of-State Entrepreneurs

One of the first out-of-staters to see the potential was James McAndrews, an Iowa cattle feeder, who in 1956 was running cattle in several states and came to the Panhandle to buy feeder cattle. He liked what he saw, so he moved his family to Hereford, bought the 1,000-head Western Feedyards and expanded it to 2,500 capacity. (Western was built about five years earlier by oilman Homer Gogswell but was empty.)

Beef magazine

James McAndrews (right) of Iowa was one of the first to see the new potential in Texas. So in 1956, he moved to Hereford and started feeding. Barry Cox (left) became his feedyard manager.

Farm Journal

Paul Engler (left), a Nebraska feeder, moved to Hereford in 1960, built Hereford Feed Yards (with a group of businessmen) and started an era. Shown with the author (right), 1967.

Living back in Iowa in 1992, the 74-year-old McAndrews recalls, "When I arrived, there were 276 head of cattle on feed in Deaf Smith County. . . . What I saw and liked was (1) the sunshine; (2) the hard soil; (3) the low rainfall, half that of Iowa; (4) the cool summer winds; (5) transportation on the Santa Fe; and (6) a world of grain."

"Only two more things were needed," McAndrews told his new banker, Dale Smiley with the First National Bank of Amarillo, "more cattle feeders and more packers. And these will soon come – in that order."

How right he was. Among the next feeders to arrive were Paul Engler and Jack Carrothers.

Engler, a 1951 graduate of the University of Nebraska, had been feeding cattle under the tutelage of Lewis Dinklage of Wisner, Nebraska from 1955 to 1960. They were feeding 25,000 to 30,000 head of cattle in 13 different locations – mostly in partnership with farmer-feeders.

Basic Raw Materials Were Here

While visiting Hereford in 1960, Engler saw two basic raw materials – grain sorghum and cattle – loaded onto the same train to be shipped to California. "That wasn't logical to me," he recalls, "and I wondered out loud why the milo wasn't fed to cattle here in Hereford." Joe Reinauer, a local businessman, gave him the answer: "We're just waiting for someone such as you to show us how."

Soon afterward, Engler was hired to supervise construction and manage the Hereford Feed Yards. Joe Reinauer put an investor group together to build the new yard. "Ten men raised $200,000," Engler says. "They contributed the capital because we couldn't borrow money then. A feedyard was a single-purpose business that could not be used for anything else, and none of the banks would consider a loan." They did finally receive a $180,000 loan from the Small Business Administration.

After the feedyard was opened, Engler recalls, he had trouble finding customers.

Many Entrepreneurs Came from Out-Of-State:

Robert Allen from Colorado

Wayne Bedwell from California

Bob Bliss from Iowa

Ike Brandon from Alabama

Melvin Cordray from Montana

Jay Crofoot from Kansas

George Feaster from California

Dee Follis from Kansas

Bob Hayes from Arizona

Lee Hicks from California

Dick Jagels from Arizona

Charlie Koontz from California

Hart Kornegay from Arizona

Wayne Owen from Arkansas

Louis Reeder from Arkansas

Milton Rudder from Arkansas

Howard Simpson from California

Bill Thorne from Missouri

W. R. (Bud) Thurber from Arizona

Hank Winter from California

Photo by Charles E. Ball

Grain farmers formed corporations and continued to build feedyards into the 1970s. From left to right, Don McLain, Carl Archer, Lloyd Buzzard and Dan Archer built Hansford County Feedyard in 1970.

"The idea of buying cattle and having someone else feed them hadn't yet been widely accepted. Not even the feedyard's investors were willing to take the risk at first."

Dinklage, however, bought a few pens of local cattle and started the yard on the successful course that it has followed since its opening in September 1961. Starting with 5,000 head, the Hereford Feed Yards continued building until reaching its present capacity of 50,000 head.

Beginning of the Cattle Feeding Era

The building of the Hereford Feed Yards started an era. While there was a general expansion among existing lots, it was not until 1962 that the "Age of the Giant Feedyards," as it was named by newspaper and magazine writers, began. One of the first "giants" was Friona Feedyard at Friona.

A. L. Black, a local farmer and cattleman, in 1961 put together a group of investors – eight farmers and cattlemen, each of whom put in $33,000. "The next thing we did was go to Colorado and hire Jack Carrothers as manager," says Black. "While he handled the cattle and helped others build new feedyards, I supervised the construction. It seemed that we never stopped building." When Black retired as Chairman of Friona Industries, Inc. in 1981, FI had 120,000 head on feed and owned six subsidiary companies.

"The idea of building a yard large enough for 30,000 head was pretty wild and big," said Jack Carrothers (in 1977), first manager of the feedyard and later president of Friona Industries. "Texas had little experience with large lots in 1962, and to begin with that many cattle on feed was unprecedented."

But the Friona Feedyard became a model for the area. Its $150,000 mill was built in California, shipped to Friona and put together there. It was a prototype for other feedyard mills in the Panhandle and surrounding states. "Our investment was $22 per head for the 30,000-head capacity yard," Carrothers recalled.

More Out-of-State Expertise

Other men moved to Texas to provide the expertise that was so necessary to cattle feeding. Robert Allen moved from Colorado, Ed Barrett from Kansas, Bob Bliss from Iowa and Illinois, Ike Brandon from Alabama, Jack Carrothers from Colorado, Melvin Cordray from Montana, E. C. Crofoot from Kansas, Jay Crofoot from Kansas, Dee Follis from Kansas, Bob Hayes from Arizona, Tom Herrick from Colorado, Dick Jagels from Arizona, Leon Miller from Colorado and California, Wayne Owen from Arkansas, Louis Reeder from Arkansas, John Richter from Washington, Milton Rudder from Arkansas, Bill Thorne from Missouri, W. R. (Bud) Thurber from Arizona, Leo Vermedahl from Minnesota.

Several were attracted from the established commercial feedyards of California. Among them: Wayne Bedwell, Mike Bowles, George Feaster, Jack Gillis, Pitts Harrison, Lee Hicks, Charlie Koontz, C. C. (Cap) Naegle, Kem Rogers, Howard Simpson, Hank Winter.

At the same time, scores of local farmers and cattlemen were beginning to build and expand. (See Chapter 3.)

Sam Thomas, agricultural manager for Southwestern Public Service Co. in Amarillo, who was an ardent promoter of feedyards, also gave credit to Paul Engler and

Photo by Charles E. Ball

Local farmers became commercial cattle feeders, too – like Grady Shepard (right) and son Pat of Hale Center. Grady Shepard received the first TSCRA Southwestern Cattle Feeder Award in 1962.

A. L. Black of Friona, Texas, led a group of investors to build Friona Feedyard in 1963, then in 1967 led a committee that enticed MBP (now Excel) to locate in Friona.

Jack Carrothers for promoting the area. "Engler and his activities attracted much glamour to the industry that it had not had before," said Thomas in 1972. "He was high profile and got a lot of national publicity.

"Carrothers also was an effective spokesman for the industry," added Thomas. "He helped others build feedyards and took the lead in organizing them into an association."

Between 1961 and 1969, growth in feedyards in the Panhandle and surrounding areas averaged 20 to 30 percent per year, overloading the area's agricultural financing houses.

Outside Capital Needed

Banks in the Panhandle had been providing some intermediate and short-term credit to build feedyards, but as Pat Malone of the First National Bank of Amarillo observed in 1969: "We feel this is outside the commercial bank's realm of activity. With the number of lots going in now, and the number still on the drawing board, plus the tremendous amount of money it takes daily, the money should come from long-term lenders – insurance companies and government agencies." In the early stages of the feedyard buildup, many loans came from the Small Business Administration and Production Credit Associations. But not enough.

So Henry Sears, president of the First National Bank of Hereford, hit the road looking for outside capital. He became well-known in New York, Chicago and Los Angeles banking circles, where he convinced many bankers of the reliability of cattle paper.

Area banks soon got into the act. By 1970, the First National Bank of Lubbock had feedyard loans exceeding $80 million. And by 1973, the First National Bank of Amarillo, with cattle loans in excess of $100 million, had become the largest bank in Texas agricultural loans.

Financing a feedyard was an expensive venture. "Cost of the mill, pens, watering system, land and some basic equipment runs about $40 per head of capacity," Pat Malone said in 1969. "A 30,000 head yard would cost a minimum of $1.2 million. And that's just the plant. . . . Once the operation gets underway, you need another

Big capital was needed. So Henry Sears, President of First National Bank of Hereford, hit the road calling on financial houses, promoting cattle feeding. Ca. 1965.

Cattle clubs and 4-H Clubs had "the fever." Allen McCloy (seated) and brother Bill won a State 4-H contest with their demonstration, "Feedlot Profits and Losses," in 1964.

$400,000 to $500,000 operating loan to buy feed, pay the overhead, etc. At 50 cents per head per day for feed, and 25,000 head in the yard, feed costs run $12,500 a day." (In 1992, those figures are more like $1.50 per head per day for feed or a total of $37,500 per day.)

By 1975, construction costs exceeded $100 per head capacity, even though the building of new feedyards had tapered off. And other money needs increased to accommodate the 1.5 million head of cattle on feed that year in Texas.

The insurance companies, which had been leaders in irrigation and farm and ranch financing, were cautious in lending money for feedyard construction because of the single purpose classifications. As the industry developed and profited, however, they and other long-term lenders did provide financing.

The search for additional funds was – and is – one of the constants of the feedyard industry. To keep their pens filled during the boom years, feedyard owners came up with some dazzling deals for raising money. Many borrowed fund raising techniques from the oil industry – investment funds, limited partnerships, management companies, feeding clubs, Sub-Chapter S corporations and co-ops.

Cattle Clubs, Investors Pool Funds

So exciting and apparently lucrative was the boom that many doctors, lawyers, businessmen and housewives wanted a piece of the "money machine." That's where the feeding clubs or cattle clubs came in. It was a way for small investors to pool their funds, spread the risk and increase volume. There were hundreds of them, reportedly over 100 in Hereford, Texas, alone.

Cattle clubs were organized in several ways – informal partnerships of three or four people, confirmed with a handshake; formal partnerships, drawn up by an attorney; limited partnerships, where one partner was not responsible for the others' obligations; and corporations. "Ten members or less can incorporate under Sub-Chapter S and be taxed as individuals," explained Dick Jagels, assistant manager of Hereford Feed Yards in 1971, who was a member of two such groups.

Youth groups had "the fever," too. In 1964, Allen and Bill McCloy of Morse, Texas won the Farm and Ranch Management Contest at the State 4-H Roundup with their demonstration, "Feedlot Profits and Losses." Their parents were Mr. and Mrs. Wilson McCloy, who built their first feedyard in 1953, with a capacity of 500 head, and by 1964 were up to 2,000. (By the 1980s, the family-owned yard held 12,000 head and Allen was the manager.)

In the 1964 demonstration, Allen described a pen of 150 steers. "When they were put on feed, January 15, the purchase price was 22 cents per pound," he said. "The cost of gain was 23 cents. . . and the breakeven is 22.4 cents per pound. . . . We need a margin of profit of 1 cent above the breakeven for it to be an economically profitable unit of our farming operation."

Foreseeing a shortage of managers ahead, Texas A&M University in 1968 introduced a Master of Agriculture degree in feedlot management. Then Texas Tech and West Texas State University started feedlot management courses.

Warehouse receipts, another innovation during the boom, made cattle feeding more attractive to out-of-state investors and to lenders, too. That was a system of inventory guarantee previously used for cotton, grain and other commodities.

Some Went Public

Eventually some feedyards – ProChemco, Western Beef, Stratford of Texas and Friona Industries – went public and sold stock.

There were mergers and acquisitions. Grain companies moved into the feedyard business to diversify and protect their grain markets – Pitman Grain in 1965, followed by Producers Grain Corporation and Prairie Grain Company in 1969, Cargill, Inc. in 1974, Continental Grain in 1975, Bartlett Grain Company in 1977.

Suddenly, the Plains of Texas, Oklahoma and New Mexico had a multi-billion dollar industry, the likes of which had not been seen since the oil-boom days of the 1930s.

But new problems were on the horizon – that would soon become painfully apparent.

ORIGIN OF CATTLE FUTURES

For better or worse, depending on your position, the idea of live cattle futures originated in Texas.

In 1962, when the fed cattle market was in a downturn, J.D. Sartwelle, president of Port City Stockyards in Houston and an innovative market man, was looking for "some way to avoid the cyclical losses." With a glimmering concept of cattle futures, he took the idea to a broker friend in Houston, who said, "I don't know, but the president of the Chicago Mercantile Exchange (CME) will be here next week to investigate a proposed shrimp contract. Why not ask him?"

So Sartwelle had dinner with Everett B. Harris, president of the CME, which also was having financial difficulties, and he was intrigued. "We desperately need new products to survive," responded Harris. "Please refine this idea and come present it to our Board."

"During the next 12 months, Harris and I attended every cattlemen's meeting and cocktail party we could – explaining, promoting, trying to determine the possible interest," says Sartwelle. "Most cattlemen thought this was pretty weird – selling cattle before they were fed or even before they were purchased."

About the same time, Harris was also approached by a representative of the Mellon Bank in Pittsburgh, who was looking for a way to protect their cattle loans. Packers were enamored with the idea, but favored a beef carcass contract.

Sartwelle held his ground. "Cattle feeders will use a live contract but not a beef contract," he argued. "And there are a lot more cattle feeders than packers."

So on November 30, 1964, the new Live Cattle Contract was offered. The open was $24, the close $23.67 and the volume 191 contracts.

Twenty-five years later, November 30, 1989, the open was $75.25, the close $74.97 and the volume 13,142 contracts.

Beef made news in 1973 – most of it negative – in virtually every newspaper and magazine in the country.

Chapter 7

THE WRECK

Inflation, higher grain, consumer boycotts, and more cattle cut the rug from under cattlemen in 1973.

The feedlot boom in the Panhandle entered the 1970s in high gear. Between 1960 and 1973, the number of cattle fed annually in Texas increased from 477,000 to 4.4 million. In 1973, there were 80 feedyards in and around the Panhandle with capacities of 16,000 head or over, 32 of them with capacity for 32,000 or more.

A long way from Texas, however, trouble was brewing. The Arabs decided to boycott oil shipments to the U.S., touching off an unprecedented spiral of inflation and concurrent recession. Russia missed its 1971 and 1972 wheat crops, then bought up the bulk of American grain stocks and so started the price of grain – and cost of feedlot gain – climbing.

A Predictable Problem

At the same time, a more predictable problem was occurring in the U.S. The cattle population had been climbing since 1967, and by 1973 had increased almost 10 percent over the start of the cycle. Between 1966 and 1974, beef production in the U.S. increased 17 percent.

The three together – inflation, higher grain and more cattle – cut the rug from under cattlemen in 1973. The result has been dubbed by cattlemen, "The Wreck." For the entire cattle industry, it was the most painful period in modern history. The cattle feeding segment was first and hardest hit.

Meat Boycotts Across the Country

The first sign of trouble came in the spring of 1973. Meat boycotts began springing up spontaneously across the country. Behind them were consumer groups with such acronyms as FIT, SCIMP, WART, HELP, HARP and National Consumers United. All told, more than 30 such protest organizations (mostly temporary) came into existence in the United States.

What the groups wanted was cheaper beef, an idea that sat anything but well on the High Plains. Responding to a resolution by Cleveland, Ohio city councilmen proclaiming April (1973) as "Boycott Beef Month," TCFA President Tom Herrick sent the following telegram:

"Noting your concern about high beef prices, the Texas Cattle Feeders Association invites you to become a business partner with us by investing in a pen of 200 head of

AP / Wide World Photos

In 1973, protest organizations sprang up, beef boycotts were organized and President Richard Nixon imposed a price freeze on beef.

cattle. First, you should be aware of the risk involved. A 650-lb. feeder steer at today's price of 55 cents per pound will cost $357.50. The steer normally would stay on feed 150 days and gain about 400 lbs. The 400 lbs. of gain will be charged at actual cost, about 30 cents per pound. Normal death loss is 1 percent and interest on your investment will run 8 percent. Thus, when the steers weigh 1,050 lbs. and are ready to sell, our breakeven cost would be $495 or 47 cents per pound. All profits or losses will be shared equally. So if the cattle sell for 52 cents per pound, our profit will be $10,500. On the other hand, if they sell for 42 cents, the loss will be $10,500."

Following the Beef Boycott in early April 1973, the TCFA Executive Committee (Tom Herrick, Glenn Deen, O. J. Barron, George Ellis, Bob Carter, Jack Carrothers, Jim Davidson) asked each TCFA member to contribute $1 for each animal on hand to an Emergency Fund to tell the cattlemen's story to the public. (See box on next page.)

Within three weeks, feeders contributed $519,350. Of this, $200,000 was forwarded to the National Live Stock and Meat Board for an expanded public relations program; and $100,000 was sent to the American National Cattlemen's Association for expanded public information and legislative programs. An additional $119,350 was earmarked for the Meat Board and/or ANCA; and $100,000 was allocated for expanded TCFA programs.

It was a letter that changed the cattle industry, because the funds generated – then and in the two years that followed – eventually revitalized the Meat Board and ANCA, as well as TCFA.

Price Freeze on Beef

But consumers – and the government – were not impressed. In June 1973, the Cost of Living Council declared a price freeze on beef at the retail level. TCFA exploded, as did most cattlemen.

Nonetheless, the freeze stayed in effect, and before summer was far along, the industry began to glimpse what it suspected was a ray of hope: When the price freeze came off, cattlemen reasoned, prices would go up. Way up, most likely.

There was good reason to think that. Retailers, though faced with steadily increasing losses on every pound of beef they sold under the freeze, continued to clamor for more of it. They paid whatever they had to, sold for what they could – some on the black market.

As the summer progressed, producers became more and more reluctant to sell. Thousands of market-ready cattle were held back, and in August, while they ate $5 grain, packers were shutting their plant doors because they couldn't buy cattle. What cattle did sell traded at unprecedented prices – up to 58 cents per pound for fed cattle and 80 cents or more for feeder steers.

LETTER THAT TURNED AROUND THREE ORGANIZATIONS

April 7, 1973

TO ALL TCFA FEEDYARD MEMBERS:

First of all, my apologies for this mimeographed letter, but the subject is urgent and this was necessary in order to get the message out this weekend.

In response to requests by many of our members, the TCFA Executive Committee held an emergency meeting today and agreed that we must take immediate action to (1) avoid a price rollback and (2) tell our story to the public and to politicians.

To accomplish this, we need your immediate support in two ways:

1. We urge you to contact each of your customers, including those in cities and in other states, asking them to call or wire their congressman and express opposition to the proposed January 10 price rollback, as proposed by the House Banking Committee. Also, ask your customers to solicit help from their bankers in contacting congressmen. We must act quickly on this, since the House of Representatives is scheduled to vote on it Thursday, April 12.

2. To tell our story – preferably on national TV during prime time – is going to take big money. So we are asking you (and all feedyards in Texas) to send $1 for each animal on hand to the Texas Cattle Feeders Emergency Fund. We need this by return mail and absolutely no later than April 13 in order to plan our message and contract for TV time.

If prices are rolled back to January 10 levels, it will mean a loss of $50 to $60 per head from today's prices. (Compared to this potential loss, we believe most owners will be happy to contribute the $1 head.) This loss will back up to every stocker calf and every cow in the country. Applied to 122 million cows and calves in the U.S., that means a $6 billion loss! Which would be a crippling blow not only to the beef industry but to the entire economy.

So we are soliciting similar support from the Texas and Southwestern Cattle Raisers Association and from every other beef association in the nation. We visualize that this money should be concentrated and channeled through some professional organization (probably the Beef Industry Council of the National Live Stock and Meat Board) that knows how to get the proper message told.

The situation is critical. And time is short. So please send your check today – $1 for each animal on hand – to the Texas Cattle Feeders Emergency Fund, 2915 South Georgia, Amarillo, TX 79109.

If you have questions, don't hesitate to call any of the Executive Committee, whose telephone numbers are listed below.

Thanks.

Sincerely,

Tom Herrick
President

[Editor's Note: Within three weeks, this letter had raised $519,350, which was divided between the Meat Board, ANCA and TCFA. Observers say it "turned around" all three organizations.]

Photo by Ron Marlow

TCFA Emergency Fund was shared with ANCA and the Meat Board. TCFA President Tom Herrick (left) presents a $100,000 check to ANCA Executive Vice President George Spencer, July 1973.

True to his promise, President Richard Nixon lifted the price freeze on September 9, 1973. Cattle which couldn't be bought for any price the day before were suddenly for sale. And the market collapsed. Between August and December, the market dropped 36 percent – from 58 cents to 37 cents per pound for fed cattle.

At the same time, grain prices were shooting upward – from $3.05 per cwt. for milo in January 1973 to $5.10 in August. Cost of gain in Panhandle feedyards, historically in the 30-cent range, jumped to 50 cents per pound and higher.

Cattle Lost $200 Per Head

The upshot of it all was that during the last few months of 1973, many of those pre-freeze 80-cent feeders, which had a 50-cent cost of gain, were marketed as 38-cent fed cattle and lost $200 per head or more.

An industry built around low-per-head profits, fast turnovers and big volumes had suddenly turned into one of fast turnovers, big volumes and ruinous losses. Through the boom years, a profit of $20 per head was considered acceptable. It didn't take many $200 losses to wipe out years of profit .

During much of 1974, the feeding industry – intent on making back those first big losses – bore the cross for the entire industry. Feeder cattle prices, bid up by still-hopeful cattle feeders, continued relatively strong into the summer of 1974.

When calf prices did collapse – after feedyards and their bankers gave up – packers started buying feeder cattle as slaughter cattle. The result was more downward pressure on fed cattle prices and still more problems for the feeding industry.

Summarizing the effect of the price drop on cattle feeders following 1973, Joe Wood, TCFA market analyst, said in October 1977, "Of the past 49 months, 32 have been loss months, with losses averaging $18.36 per head. So if a cattleman had stayed with the feeding business from 1973 to 1977 and fed 200 cattle per month, he would have lost $179,928.

"Had he been marginally financed with only enough capital in the beginning to

Photo by Joe Monroe

Tough times for everybody in the business. "Of the past 49 months, 32 have been loss months," said TCFA Market Analyst Joe Wood (above) in October 1977.

Consistent losses caused cattle-on-feed numbers to drop 54 percent, from November 1973 to March 1975, forcing a dozen Texas Panhandle feedyards to close their gates (right) in bankruptcy.

meet minimum requirements – 35 percent of the original cost of a calf – he would have been wiped out totally by mid-1974."

Many Feeders Wiped Out

Many feeders were indeed wiped out, and their absence was tough on the feedyards. During much of 1974 and most of 1975, pen space went wanting for occupants. Trying to take advantage of low calf prices and to minimize high feed costs, the feeders who stayed in were placing larger calves by 1974, and the result was that what cattle were placed didn't stay as long as before. Many feedyards dropped to 60 percent occupancy, and during the pits a number had only 30 percent of their pens filled. Between November 1973 and March 1975, the number of cattle on feed in Texas dropped 54%. The Texas Panhandle lost almost a dozen feedyards to bankruptcy during the 1973-75 debacle and its aftermath.

The wonder is not, however, that so many feedyards were lost, but that so many survived. It was a baptism of fire, "The Wreck," a coming of age for an industry which had grown fast and big.

Photo by Ralph Leone

Located on busy Interstate 40 in Amarillo, TCFA flies the colors for travelers from coast to coast.

Chapter 8

A New Association

The need became apparent for an association to service feeders and help guide their future.

By the mid-1960s, the new Texas cattle feeding industry – a rapidly growing young giant – was feeling the need for an association to service the feeders and help guide their future. Out of that need was born the Texas Cattle Feeders Association, in 1967.

Prior to this time, some of the early feeders were associated with the Cattle Feeders Division, established in 1961 by the Texas and Southwestern Cattle Raisers Association (TSCRA), with headquarters in Fort Worth. President Dolph Briscoe saw new feedyards emerging all over the state. Cattle on feed in April 1961 showed a 35 percent increase over the previous year – up to 218,000 head. So Briscoe appointed Norman Moser of De Kalb as chairman of the first TSCRA Cattle Feeders Committee.

The first thing the new committee did was hire a director, Lloyd Bergsma, who had been Extension Livestock Marketing Specialist at Texas A&M University, and change their status from a committee to a division of TSCRA. One of the early services established by Bergsma was a market information network involving 30 feedyards. Also, he called on USDA to issue a monthly Cattle on Feed Report for Texas (rather than quarterly), which it did in 1964.

Cattle Feeders Wanted Autonomy

As cattle feeding grew in Texas – 1.4 million were marketed in 1966 – and the feedyards tended to concentrate in the Panhandle, the feeders in that area began to want more specialized and more localized services. Says Bob Carter of Plainview, who started feeding in 1955 and who by 1966 was a director of the TSCRA Cattle Feeders Division: "We wanted more autonomy; we wanted to elect our own directors, rather than have them appointed by the TSCRA president; we wanted to set our own dues; and we wanted an office in the Panhandle."

So a group of feeders – Bob Carter, Jack Carrothers and Robert Allen among them – met in Plainview, chartered a plane and flew to Dallas to meet with TSCRA President Ben Carpenter. "He wasn't very understanding of our desires," says Carter, "so on the way home we decided to pursue a new association."

The viewpoint of TSCRA was that they had spent a lot of effort and money helping feeders; there was an obvious potential for members and revenues; and they were reluctant to give it up.

Organized May 18, 1967

Nevertheless, on May 18, 1967, ten cattle feeders met at the Holiday Inn West in Amarillo to discuss the possibility of forming a new association. Attending were Jack Carrothers, Friona; Robert Allen, Tulia; J. O. Parker, Happy; Richard Jagels, Hereford; R. M. (Bob) Carter, Plainview; H. D. King, Muleshoe; Grady Shepard, Hale Center; Walter Lasley, Stratford; Rex McAnelly, Pampa; and Gene Newman, San Angelo.

Not all of the Texas feeders were in favor of forming a separate association. A questionnaire had been mailed to feeders before the meeting, and Carrothers reported 63 of the 81 respondents desired a new association.

A 1970 executive committee meeting. From left: Robert Allen, Glenn Deen, Joe Easley, Bob Carter, George Ellis, Jack Carrothers, and Tom Herrick.

The feeders at the meeting voted seven to three in favor of establishing a new association and elected Jack Carrothers president, R. M. (Bob) Carter vice president and Robert Allen treasurer. King, Jagels, Lasley and Parker were named to a temporary executive committee.

The first board of directors consisted of O. J. Barron, Jr., Spur; Tom Bratcher, Dallas; Jim Davidson, Sudan; Glenn Deen, Dumas; Joe Easley, Hereford; Tom Herrick, Amarillo; John Jarrell, Snyder; E. George Luckey, San Antonio; S. B. Middlebrook, Vernon; C. C. Naegle, Friona; John Pitts, Jr., Perryton; M. B. Powell, Haskell; Tom B. Simmons, Jr., Lubbock; and Johnny F. Summerour, Dalhart.

Other action taken at the May 18, 1967 meeting: Texas Cattle Feeders Association was selected as the name for the new organization; Lloyd Bergsma, who had been director of the Cattle Feeders Division of TSCRA, was hired as the first executive secretary of the Association; and James W. Witherspoon of Hereford was selected as association attorney.

James W. Witherspoon, TCFA attorney, 1967, and organizer of several feedyards.

Dues $25 Plus 10 Cents Per Head

To finance the new Texas Cattle Feeders Association, members of the executive committee pledged $1,000 each, of which $500 was made available immediately. Other members of the board were asked to assist with interim financing. At the first board meeting on May 31, 1967, dues for individuals and firms engaged in cattle feeding were set at $25 plus an assessment of 10 cents per head for all cattle fed. Associate memberships at $100 were provided for individuals and firms allied with the cattle feeding industry but not actually engaged in cattle feeding.

When the first quarterly meeting of the board was held in Amarillo, July 20, 1967, there were 43 regular members and 36 associate members in the Association and $4,593 in the bank. Five standing committees were established: Research and University; Membership and Finance; Insurance; Water and Air Pollution; and Technical.

The new association grew rapidly. By the second quarterly meeting, October 19, 1967, there were 72 regular members and 40 associate members. The budget established for the fiscal year beginning October 1, 1967, was $40,675.

Back at the Cattle Feeders Division of TSCRA, Dudley Campbell was hired to replace Lloyd Bergsma. Campbell was well qualified, having been on the staff of the American National Cattlemen's Association and the National Live Stock and Meat Board. But the Cattle Feeders Division struggled in competition with the new association and folded in February 1968.

Rejected By ANCA

Before it folded, however, some cattle feeders were divided in organizational loyalty and deep wounds developed between TSCRA and TCFA, wounds which took several years to heal. At the January 1968 convention of the American National Cattlemen's Association (ANCA) in Oklahoma City, a delegation of TCFA officers and directors appeared for membership in ANCA. They had assurances from President John Guthrie and Executive Vice President Bill McMillan that they would be accepted. Officers of TSCRA protested, however, and prevailed on the ANCA board to vote against admission of the new TCFA. So the TCFA delegation stormed out of the meeting!

A few months later, however, following a change in leadership at TSCRA and ANCA, the new association (TCFA) was accepted and has been a strong member of ANCA (now NCA) ever since.

T-BONE CLUB, FORERUNNER OF TCFA

Every organization begins somewhere, and it was the T-Bone Club from whence came TCFA.

As early as 1964 – three years before TCFA – a few visionaries could see the need for a rallying organization to coordinate and serve the emerging cattle feeding industry on the Southern Plains. So the T-Bone Club was established, primarily by Sam Thomas, Manager of Ag Development, Southwestern Public Service; Don Hileman, Manager of Amarillo Chamber of Commerce; Bill Nelson, Executive Vice President of Grain Sorghum Producers Association; Walter Nichols, Ag Manager of Pioneer Natural Gas; Jack Tompkins, Ag Director of KVII-TV; and Don Jones, cattle buyer for Wilson & Co.

The purpose of the organization – originally formalized with constitution, bylaws and membership dues – was to provide an occasion for feeders and allied industry people to meet once a quarter to exchange ideas and hear a technical program. To keep it from being all work, however, an "attitude adjustment hour" was made a part of each meeting. Later, it became an informal, loose-knit gathering.

The project was a success from the start. People attended from throughout the Panhandle. Some of the memorable speakers included Dolph Briscoe, candidate for Governor of Texas; John Trotman, President of ANCA; Dr. Bart Carden, President of Arizona Feeds; Allen Grant, President of American Farm Bureau Federation; U.S. Rep. Tom Foley, House Agriculture Committee; Texas Rep. Bill Clayton, Speaker of the House; Clancy Adamy, President of the National Association of Food Chains; Eddie Collins, Iowa market news commentator; Ace Reid, cartoonist; and many more.

The T-Bone had five well-known presidents: Jim Davidson of Sudan, Walter Lasley of Stratford, Jack Carrothers of Friona, Pat Shepard of Hale Center and Rex McAnelly of Pampa. All later became directors of TCFA.

After TCFA was organized in 1967, the T-Bone continued for a few years. But as TCFA sponsored more and more educational seminars, the only justification for T-Bone was social, and the last meeting was in 1981.

Lloyd Bergsma, TCFA's first executive officer (left), greets members at the 1970 TCFA Convention in Amarillo, with Cap and Sue Naegle.

Membership grew rapidly and by 1982 had reached 6,000 members, supporting a $1.8 million budget. Membership Director Jim Peterson (right) with Secretary Charis Young.

Membership Mushroomed

From then on, TCFA membership mushroomed, as did the Association's budget. There were 390 members by the end of 1968, and the number increased nearly every year until 1984 when it peaked at 7,128 and later plateaued at around 6,500. The budget grew also – to $312,000 in 1972, to $1.8 million in 1982, to $2.8 million in 1992.

South Texas feedyards, although a remote 500 to 600 miles from the Association headquarters, have been an important part of that growth and have been given specific attention. In 1973, for example, Membership Director Mike Riethmayer was stationed in San Antonio. And in 1982, TCFA commissioned Texas A&M researchers to conduct a study, "Future Alternatives for South Texas Cattle Feeding and Beef Marketing Systems."

Allied industries also have been an integral part of the growth. Over 500 Industry Associates pay $100 per year in dues and assist TCFA in numerous ways, both financial (over $125,000 in 1991) and in-kind help. Examples: The Industry Associates help plan and finance the TCFA Annual Convention, educational seminars, the TCFA Education Foundation . . . and much more. The chairman of the Allied Industries Committee is a member of the Board of Directors.

To earn such spectacular growth, member services needed to constantly increase. And increase they did!

Market Information. From the beginning, market information has been a valued service to members, because it helped put "dollars in the pocket" every time cattle were sold. TCFA started with one market reporter in 1967 and, by 1987, the Market Department had grown to five people. The stated objective always has been, "To keep the seller as well informed as the buyer."

Many managers consider this information so valuable that they won't sell cattle until first calling TCFA to see what the market is "at this moment," not what it was

Elmo, a 1978 "spokesman" for TCFA, created by cartoonist Jerry Palen.

yesterday or last week. Further attesting to its effectiveness, some packer buyers, when paying a top price for the day, prevail on the manager not to report it to TCFA.

TCFA has had 20 market reporters/analysts: Cathryn Covey, Charles Ray Delay, Garland Weeks, David Hales, Gery Donnell, Joe Wood, Jim Gill, John Kuker, Robert Price, Chris Hyndman, Thomas Horndt, J. B. Douglas, Larry Fullen, Stan Price, Steve Johnson, Robert Hughes, Patti Adams, Mike Boswell, Brent Koller and Jerry McCarty.

Through the years, USDA has traditionally collected cattle on feed data for its official reports. By 1980, the USDA recognized that TCFA was getting more cooperation from Texas feedyards, thus more data, than USDA was. So the USDA contracted with TCFA to collect cattle on feed information, a contract which remains in force in the 1990s.

In 1987, TCFA took a bold step toward expanding its market information – in addition to telephone and Commodity News Service quotes and news – to a satellite-computer network, called "M-Net." The system was implemented by Bill Carrothers, son of the first TCFA president, and later managed by Steve Johnson and Tom McDonald. TCFA furnishes each participating feedyard with a computer. So when a feedyard makes a sale, it enters the pertinent information into the computer and relays it via satellite to TCFA. There the information is consolidated with sales information from other feedyards and automatically sent back to M-Net feedyards every 30 minutes, or more frequently if market conditions dictate.

Legislative Representation. Since legislative affairs always has been a priority for the Association, TCFA officers and staff have traveled to Austin and Washington frequently. Since 1977, the entire Board of Directors has gone to these capitals, also, to lobby and present their ideas. They've hosted many luncheons or dinners, where TCFA steaks were served, and have taken a box of steaks to each senator and congressman. This gets their attention! (State legislative affairs have been handled by the Oklahoma Cattlemen's Association in Oklahoma City and the New Mexico Cattle Growers Association in Santa Fe.)

To supplement this legislative effort, TCFA in 1976 formed a political action committee, Beef-PAC, which Brenda Higley serves as treasurer. (See page 111.)

Photo by Ralph Leone

Timely market information, one reason for forming TCFA, was the first service activity. Market analysts in 1982 were (from left) J.B. Douglas; Jim Gill, market director; and Tommy Horndt.

Beef promotion became the second TCFA activity, after market information. In 1969, TCFA joined with the Texas Department of Agriculture in sponsoring billboards. Jack Carrothers, TCFA president (left) and John C. White, Texas commissioner of agriculture.

Beefiestas, in Dallas and Atlanta, 1970-71, drew food editors and dieticians. Sponsored by TCFA and National Beef Industry Council.

TCFA hired Ross Wilson, a former congressional aide, as a fulltime government affairs director in 1985 and reinforced his efforts with two well-known Washington lobbyists – C. W. (Bill) McMillan, a former assistant secretary of agriculture and former CEO of NCA, and Denise Bode, a tax attorney.

A few of the successful legislative efforts, which TCFA instigated or helped to obtain, usually in concert with NCA and other associations, include:

- A Veterinary Diagnostic Laboratory in Amarillo, 1975, which serves the three-state area.
- Prompt Payment Legislation, 1976, requiring packers to pay for cattle by the close of the next business day following purchase.
- Beef Research and Information Act, 1975, which eventually resulted in a $1-per-head checkoff that became effective in 1988.
- Counter-Cyclical Meat Import Act, 1979, which reduces beef imports when domestic production is up.
- Ad valorem taxes removed from Texas livestock, 1981, saving owners up to $4 per head per year.
- Clear title legislation, 1985, at the state and federal levels gave clear title to buyers of ag commodities and eliminated the possibility of having to pay twice on mortgaged products.

Regulatory Liaison. TCFA has from the beginning had a concern for and intense interest in protecting the environment – one of the original committees was the Water and Air Pollution Committee. To assist feedyards in dealing with the 15 agencies (state and federal) attempting to regulate feedyards, TCFA hired Dr. Richard McDonald in 1974 (Chapter 4). He and his successor, Ross Wilson, have maintained a close relationship with the agencies involved, often helping them write regulations that are acceptable to feedyards.

Beef Promotion. As early as 1971, TCFA joined hands with the Texas Department of Agriculture in promoting Texas Grain Fed Beef. Together, they put billboards near major highways, posters in supermarkets, brochures and table tents in restaurants. In

1970-71, in cooperation with the Beef Industry Council, they sponsored Beefiestas in Dallas and Atlanta, attracting newspaper and magazine food editors from seven states.

In 1974, following the beef boycott, TCFA initiated the idea of a national beef checkoff, which did not become reality ($1-per-head checkoff) until 14 years later. Always believing in national beef promotion, TCFA in 1973 forwarded $200,000 to the Beef Industry Council of the Meat Board and later increased its contribution to $250,000 per year, until the 1980s when the checkoff began.

In 1978, when beef was plentiful and cattle prices were depressed, TCFA took the lead in a Beef Blitz (intensive newspaper and radio advertising) in the East Coast states. Also that year, TCFA hired a director of nutrition, Anne Ilse Anderson, to work with dieticians, food editors, restaurants and supermarkets.

Research. Recognizing that research is the foundation on which our industry is built, TCFA has regularly planted "seed money" into research projects. Over 60 research projects have been funded at Texas A&M University, Texas Tech, West Texas State University, Oklahoma State University and elsewhere.

The Research Committee also advanced the interfirm comparison program, started earlier by Texas A&M University. Called INDEX (for INdustry Data EXchange) the program permits feedyards to compare their monthly performance with other feedyards

Infrequent TCFA Managers Roundtables – whenever a topic or problem warrants one – have been popular through the years. Ca. 1974.

in 72 different categories – rate of gain, cost of gain, death losses, feed consumption, utility cost per 1,000 head, labor cost per 1,000 head, etc.

Also in 1974, the Research Committee began a program to price grain purchases based on accurate measures of nutritional and processing values. Called the Southwest Scale of Grain Premiums and Discounts, it has become the standard of the trade. Some feedyard managers say it saves them money. But the real purpose was to reflect the true feeding value of the grain and to price it accordingly.

Insurance and Safety. In addition to group health and life insurance, TCFA in 1979 introduced a Total Insurance Plan (TIP) – covering buildings, livestock, liability, workers compensation, etc. – which offered significant savings to feedyards when tied into a feedyard safety program. Charlie Christopher, a former feedyard manager, was hired to manage the safety program and he later was succeeded by Derrell Havins.

Santa Fe Railway

Volume grain shipments in unit trains of 25-100 rail cars were initiated by TCFA in 1982, saving feedyards 10-20 cents per cwt. on grain from the Midwest.

Communications. The TCFA *Newsletter* – with a terse, interesting style – is read by over 6,000 people weekly. Another popular publication is the ***Cattle Feeders Annual***, initiated in 1971, which contains feature articles on important issues; plus directories of Feedyard Members, Industry Associates, professional services and products; plus a Beefacts section with all the beef cattle statistics that matter – in one place. In addition, the Communications Department handles public relations, brochures and special reports.

TCFA has had seven communication directors through the years – Steve Munday (1974), Glen Starcher (1975-76), John McCarroll (1976-77), Steve Cornett (1977-78), Lynn Heinze (1979), Lee Wilcox (1980-84) and Burt Rutherford (1985-).

For several years, the Communications Department serviced the Beef Promotion Committee and their activities, which took a major part of the budget. After the $1-per-head checkoff became effective in 1988, however, the promotion function was handed to the Texas BIC and the Cattlemen's Beef Board. The committee changed its name to Industry Promotion Committee and switched to promoting "Cattle Feeding Country" (Texas, Oklahoma and New Mexico) as the preferred place to feed cattle. With a budget of over $100,000 per year, this promotion is being done through booths at state and national cattlemen's association meetings, slide shows, brochures and movies.

Other TCFA Services

Among the other TCFA services are educational seminars and meetings; managers roundtables, forums for managers to discuss timely topics; the Fed Beef Challenge, a leading carcass show; an arbitration program, where members can get their differences resolved by peers, rather than endure expensive attorney and court costs; Beef-PAC, a political action committee (see page 111); and more.

One of the profitable innovations by TCFA is volume grain shipments, initiated in 1982 for feedyards to buy and ship grain from the Midwest in unit trains of 25 to 100 rail cars. The savings sometimes run 10 cents to 20 cents per cwt. and often reduces local grain costs similarly.

1900 Percent Return On Investment

In 1979, Dr. John A. Hopkin, a distinguished professor of agricultural economics at Texas A&M University, was commissioned to conduct a cost-benefit study of TCFA dues, an unprecedented study for an association. He interviewed 74 feedyard managers who estimated that for each 50 cents cost in dues they received $9.51 in benefits. A return of 1900 percent!

Photo by Ralph Leone

A high level of professionalism by staff gained for TCFA recognition, accomplishment, and success. Ca. 1976.

This information gave the staff a new tool with which to sell membership. And sell they did. Membership participation – actual dues collected divided by the potential on all cattle fed in the TCFA area – in 1980 increased to 52 percent. That was a new high for TCFA and "unbelievable" in the eyes of most cattlemen's associations. Actual dues collected were $1,449,916 – also a new record for TCFA or any other state cattlemen's association.

The upshot was that the Board of Directors decided in 1980 to reduce dues 20 percent – from 50 cents per head to 40 cents. That, too, was unprecedented in the association world – no association was known to have ever reduced dues out of strength.

In agreeing to reduce dues, the Board of Directors also voted that each Feedyard Member must sign a membership agreement to, among other things, pay dues on 100 percent of the cattle fed. That, too, increased membership participation – to 68 percent in 1981 and eventually to 82 percent in 1989. Another new record for TCFA!

Five membership directors helped make this growth happen: Bud Patterson (1970-73), Mike Riethmayer (1973), Gary Culp (1974-77), Jim Peterson (1977-84), Ron Banks (1984-87), Don Clift (1987-).

Photo by Ralph Leone

Brenda Higley, administrative assistant since 1974, has been creator and expeditor of many successful programs. Shown with the author, ca. 1982.

High Level Of Professionalism

Such progress would have been impossible without bold leadership by the volunteers and a high level of professionalism by staff. It took staff members like (to name only the senior staff who have received 15-year pins): Brenda Higley, who came in 1974 and has been the "person behind the scenes" in the success of many programs. As administrative assistant to the executive vice president, she has been a confidant, a creator, an organizer, an expeditor and an inspirer. Also, she has served as staff assistant to the elected officers, the Board of Directors, Beef-PAC and the Leadership Training Program.

Photo by Ralph Leone

During its 25 years, TCFA has had three home offices: 800 West 9th Street (lower right); 2915 South Georgia (right center); and since 1980, 5501 West I-40 (top), all in Amarillo. Santa's sleigh, pulled by calves (lower left) attracts many viewers at Christmas time.

Other key, long-time staffers include: Peggy Sherron, the receptionist who came in 1973 and always makes people feel glad they called or visited. Jim Gill, the market director, who came from USDA in 1974 and helped build a unique market information department. Richard McDonald, who came in 1974 and distinguished himself in the areas of administration and government relations before being named executive vice president in 1988. Charles E. Ball was the executive vice president for 16 years, 1972-88, when much of the growth took place. (For other staff, see Appendix B.)

Since its birth in 1967, TCFA has lived at three headquarters locations in Amarillo – in rented offices at 800 West 9th Street, 1967-1971; in rented offices at 2915 South Georgia, 1971-1980; and in its own building, 5501 West I-40, since 1980. The current headquarters was built in 1980 for $500,000 – cash.

New Mexico and Oklahoma Feeders Join

Back in 1974, the Cattle Feeders Committee of the New Mexico Cattle Growers Association (NMCGA) became interested in a market information program, similar to TCFA's, but did not have enough volume to justify it. (That year, 377,000 head were marketed in New Mexico and a few feedyards were paying dues of 2 cents per head to NMCGA.) Consequently, a deal was negotiated whereby TCFA could solicit members in New Mexico and collect the regular dues. In return, TCFA would furnish all feedyards the same services enjoyed by Texas feedyards (except state legislative representation) and, in addition, rebate 10 cents per head to the Feeders Committee of NMCGA. Also, TCFA would send 20 cents of the 50 cents collected to NCA and the Beef Industry Council of the Meat Board, for national programs.

Shortly thereafter, several Oklahoma feedyards expressed interest in TCFA market information. So later in 1974, TCFA negotiated the same deal with the Oklahoma Cattlemen's Association (OCA), including a 10-cent rebate to OCA.

It was a win-win-win deal – for TCFA, for NMCGA, for OCA – and a similar arrangement has continued for 18 years.

For the 25th Anniversary celebration, which took place throughout 1992, a special logo, with "25" around the regular logo, was designed by Amarillo artist Gayle Allston.

25
TCFA
ANNIVERSARY
1967 - 1992

BEEF-PAC OPENS DOORS

Early in the life of TCFA, it became apparent that legislative clout was needed to protect and advance the interests of cattle feeders – at the state and national levels.

Cattlemen had observed how labor unions, then dairy organizations, had gained political victories by supporting candidates friendly to their causes. Yet, associations were prohibited from making political contributions – only individuals could.

So in 1976, TCFA formed a political action committee, called Beef-PAC. TCFA never had an illusion that it could, or would want to, buy votes. But it proved correct, that supporting a candidate who was friendly to our industry – or at least had the same values as cattle feeders – would pay off. And it did, primarily in that it opened doors. "Our candidates," if elected, have always been available to TCFA representatives and willing to listen to our viewpoint.

Beef-PAC was the first political action committee registered by cattlemen to support candidates at both the state and national levels. (Two associations had small state PACs and the ANCA had a relatively inactive national PAC.)

Initially, TCFA members were asked to "contribute $100, $50, $25 or whatever you can afford." Then, in 1986, the Feeders 500 Club was formed to give special recognition to members who contributed $500. And in 1989, the Century Club was added for members who contributed $100. Now most of the funds come from those two groups.

According to Brenda Higley, Beef-PAC treasurer during its 16-year history, Beef-PAC was well received the first year ($6,510 in contributions) and has grown every year since. During the 1990 election campaign, Beef-PAC contributed $189,732 to 131 candidates for state and national offices, making it the 15th largest agricultural PAC in the United States in 1991.

Photo by Tom Curl

A 1973 scabies outbreak compounded the weather problem, since Texas Animal Health Commission required all cattle to be dipped, regardless of cold weather.

Chapter 9

MAJOR ISSUES

A myriad of major issues – some took more time, more effort, more money.

Each of the past 25 years, it seemed, was filled with major issues for TCFA leaders, the members and the staff. Some issues, however, took more time, were more persisting and had more impact than others.

Worst Winter Ever

Gery Donnell, TCFA market analyst, reported 35,754 cattle died in Panhandle feedyards during "worst winter."

The scabies outbreak during 1972-73 was very frustrating and expensive for feedyards. That was called the "worst winter ever" in the Texas Panhandle, at least in recent memory of feeders. The storm started Halloween night, and the ice and snow didn't completely melt until spring.

Mud and cold played havoc with feedyard performance. But the bigger losses were on wheat pasture. TCFA solicited help from the Texas Agricultural Extension Service and its county agents to survey the extent of damages in the country, while TCFA Market Analyst Gery Donnell surveyed feedyards. The results: 35,754 head or 2.15 percent of the cattle in Panhandle feedyards died. The combined feedyard and pasture loss, according to Donnell, came to $30.5 million.

"Some heavy cattle have been standing in deep mud for days and are fatigued," said Charles E. Deyhle, DVM, consulting veterinarian at Clarendon. "When they do lie down for a few hours, paralysis develops on the side in the mud, they can't get up, fluid gets in the lungs and they die."

The Scabies Scrap

Then came scabies, the small mite that causes animals to itch, rub and lose hair. The Texas Animal Health Commission (TAHC) ordered dipping of all animals in quarantined yards – about 45 of them. That was before Co-Ral was approved, so the only treatment was toxaphene or hot lime and sulphur dip.

At that time, because of the booming demand for feeder cattle, many feedyards were buying small calves – as light as 200 pounds from Florida and other southeastern states. Many got off the trucks shivering and never stopped until they were dead. Several shipments resulted in 25 - 35 percent death loss and there were isolated reports of above 50 percent loss. Feedyards challenged the wisdom of dipping these animals, which already were stressed and suffering from shipping fever. Likewise, dipping was

Photo by Charles E. Ball

The winter of 1972-73 resulted in unprecedented losses, especially among cattle on wheat pasture (right). Rendering plants were overwhelmed with dead cattle (above). TCFA reported 35,754 died in Texas Panhandle feedyards.

Photo by Ron Marlow

hard on the heavy, fat cattle. Most dipping vats were designed for thinner feeder cattle. More than one feedyard lost big steers that got stuck in the vat and drowned.

Numerous meetings were held with TAHC officials, in Austin and Amarillo. Feeders' emotions were intense. At the insistence of TCFA, TAHC did make some concessions, but the problem was not solved until legislation was passed that reclassified scabies as a parasite rather than a disease and until Co-Ral, a systemic, was approved.

Out of this nightmare came one plus – TCFA was able to obtain legislation that provided for a Veterinary Diagnostic Laboratory in Amarillo.

Consumer Boycotts, Price Freeze

Possibly the most far-reaching issue that TCFA has had to deal with was the consumer boycotts in late 1972 and early 1973. They triggered a domino effect that put many cattlemen out of business and, in other ways, changed the nature of the industry.

The first protests by consumers were heard in late 1972, and they spread like wildfire in the following months. In January 1973, HELP (Housewives Elect Lower Prices) of St. Louis called for "Meatless Thursdays." In February, AFL-CIO President George Meany called for price controls on food (while pushing for an 8 percent increase in wages for 5.5 million workers). In March, the Rhode Island Senate passed a resolution calling for a "Meatless Day," in protest against "soaring meat prices." (By this time, Choice steers were selling for $45 per cwt. in the Panhandle, compared with $41 two months earlier and $35 six months earlier.)

The heat was on President Richard Nixon. So on March 29, 1973, he announced a price freeze. TCFA President Tom Herrick immediately issued a statement that was printed nationally: "The Texas Cattle Feeders Association feels it is unfortunate the president was pressured into placing controls on beef products. We feel it can only curtail future production of beef and lead to still higher prices. The only way to stabilize beef prices is through increased production, which will come about only if producers have sufficient economic incentive."

Nevertheless, during "April Fool's Week," a rash of boycotts took place across the nation – by about 30 groups with names like FIT, SCIMP, WART, HELP, HARP, National Consumers United and Consumers Federation of America. In response, the U.S. House Banking Committee voted to roll back food prices to January 10, 1973 levels, when Choice steers were selling for $40-$41 per cwt.

The TCFA Executive Committee hopped a plane to Washington, where they joined 100 other incensed cattlemen in an effort to put out the fire and avoid a rollback. They thought they had succeeded, but as soon as the cattlemen left town the Congress reneged and yielded to the upset consumers.

Emergency Fund

Back in Amarillo, TCFA had called on members for $1 per head for an Emergency Fund – to tell the cattlemen's story – and in three weeks raised $519,350. This, too, had far-reaching effects. The Emergency Fund was shared with NCA and the Meat Board; it turned around all three organizations, including TCFA; and eventually led to the national $1-per-head checkoff, which has financed many positive results.

But not until cattle feeders had gone through "The Wreck" (Chapter 7). Cattle feeders endured the summer of 1973, anticipating higher prices when government controls came off. Instead, prices plummeted – from $58 per cwt. in August to $37 in December. Many cattlemen lost $100 to $150 per head and some over $200.

First of Bankruptcies

Among the first of many to go bankrupt during the following year was James Louie Heller, a dealer from Illinois who was buying cattle on a large scale in the Panhandle, mostly for Iowa Beef Packers (IBP). When the checks started bouncing, TCFA called a meeting of creditors with Heller at the TCFA offices on February 16, 1974. Thirty feedyard managers and owners showed up, each due from $28,000 to $419,000.

Heller, a popular buyer, was very candid – said he was "down $2.5 to $3 million, primarily because of losses in futures during the past six months." He asked for time to work out of it and pay off the creditors. The feedyard representatives were empathetic but wouldn't agree to a delay. Instead, they elected a committee of three feeders to press for bankruptcy by Heller and to consider a suit against IBP, for whom some believed Heller was an agent.

Pat McCarthy, Amarillo, ca. 1974

On February 27, 1974, Heller was declared bankrupt and, in due time, a suit was filed against IBP. This suit was lost and appealed to the U.S. Court of Appeals, which in 1983 ruled that Heller was in fact a dealer, not an agent. Therefore, IBP was not liable for his debts.

The next big bankruptcy was American Beef Packers (ABP), which operated six plants from its Omaha, Nebraska headquarters, including a new plant at Cactus, Texas, near Dumas. This modern plant, originally scheduled to cost $14 million, ended up costing $20 million when finished in 1973, according to the company.

When ABP became slow in paying for cattle – seven to 10 days – and rumors about their solvency began to surface, TCFA called Frank West, president of ABP, several times on behalf of its members.

As late as Saturday, January 4, 1975, West assured TCFA, "The rumors are just bologna. . . . We pay for cattle within 48 hours. . . . And we're not in financial trouble." On Tuesday, only three days later, ABP declared bankruptcy, owing cattle feeders over $20 million. Of these, at least 20 were TCFA members, each due $35,000 to $600,000.

The TCFA Insurance Committee had developed a TCFA Credit Insurance Program the year before, but only about 20 feedyards had taken advantage of it. TCFA worked to keep ABP in business under Chapter 11 of the bankruptcy act, as members felt a need for that market. Swift bought the plant later in 1975 for $18 million and Monfort bought it from Swift in 1988.

Photo by Ralph Leone

C.C. "Cap" Naegle, chairman of TCFA Market Committee, pushed for prompt payment and was boycotted by packer buyers in 1974.

Prompt Payment

These two bankruptcies underscored one of the biggest issues of the 1970s for TCFA feedyards – prompt payment. As early as 1973, feedyard members were complaining about "slow pay" by packers – seven to 10 days typically and frequently up to 14 days. Although written by a local plant, the checks often were on a remote bank, like in Spokane, Washington or Bangor, Maine, where the transportation was poor and processing was slow. On a typical pen of cattle, a 10-day float could cost the owner over $100 in interest. But a bigger worry was that the feedyard or owner could lose the total value of the pen, if the packer went under during the float period.

C. C. (Cap) Naegle, the highly-respected chairman of the TCFA Market Committee during that time, was voted as the spokesman to deal with packers. In addition to prompt payment, the Market Committee and Board of Directors had developed Uniform Marketing Standards, which made the packers unhappy. Among other things, the Standards called for: (a) Weighing cattle early on the morning of delivery; no afternoon weighing. (b) A maximum of 4 percent shrink (some packers were asking for and getting 5 percent). (c) No adjustments for mud; mud should be taken into consideration in the bid price. (d) Delivery date to be no more than seven days after sale (some packers were taking two or three weeks to pick up cattle). (e) All cattle to be paid for on the day of delivery.

(Regarding shrink, one packer buyer quipped: "You set ze price, I set ze shrink!")

At Managers Roundtables on May 2, 1974 in Amarillo and May 4 in San Antonio, 58 managers signed up to support and abide by these standards. Then, on June 5 at the First National Bank of Amarillo, the Market Committee met with 20 packer representatives and 30 bankers to ask their cooperation in speeding payment.

The packers resisted. For example, Bill Nicholson, head buyer for Missouri Beef Packers (MBP), stood and declared: "Gentlemen, what you are proposing would cost

TCFA pushed for Prompt Payment Bill, which was signed by Gov. Dolph Briscoe in 1975. Standing from left: TCFA Executive Director Richard McDonald, Rep. Susan McBee of Del Rio, Sen. Grant Jones of Abilene, and Rep. Dave Finney of Fort Worth.

my company $2 million a year in lost interest. We are not going to do it." In the days that followed, Armour, ABP, Morrell and Swift also announced their opposition.

So in January 1975, TCFA went to the Texas Legislature and obtained the Texas Prompt Payment Law, which required payment the day of delivery and gave sellers a priority lien on the carcasses.

Packers became more unhappy and "black balled" Naegle, passing him by or bidding him below the market. The pressure was so great that in 1976 he felt forced to

Cattlemen listened to consumer leaders. Ellen Zawel (left) president of National Consumers Congress, spoke at the 1975 TCFA Convention. Carol Tucker Foreman, president of Consumer Federation of America, became an assistant secretary of agriculture and spoke at the 1978 TCFA Convention.

resign as chairman of the Market Committee and was replaced by Bob Carter.

Meanwhile, in 1975, Oklahoma and Kansas also passed prompt payment laws, just different enough from Texas to cause confusion. So in 1976, TCFA's Charlie Ball and attorney Wayne Sturdivant of Amarillo went to Washington to work with the American National Cattlemen's Association and the National Livestock Feeders Association in passing a national prompt payment law, as an amendment to the Packers and Stockyards Act. Sturdivant, who had advised TCFA through the ordeal with packers and in drafting the Texas law, was more familiar with the situation than anyone in Washington, so he ended up practically writing the new law. It passed in 1976, over strong opposition by the powerful American Meat Institute, which represents packers.

Cattlemen Listen to Consumers

During the same years, TCFA was attempting to understand and work with consumer organizations. "Nothing personal against cattlemen," their leaders attempted to assure cattle feeders. "The beef boycott of 1973 was not designed to be an attack on cattlemen, nor was it a planned and managed event," said Ellen Zawel, president of the National Consumers Congress, to the 1975 TCFA Convention in Fort Worth. "It was not a call for the extended and ill-conceived meat price ceiling that eventually worked to the detriment of producers and consumers."

The following year, at the 1976 TCFA Convention in Amarillo, Karen Kalish,

Producer Suits Against Retailers and Packers

In the mid-1970s, cattle feeders across the nation still were reeling from "The Wreck," when many lost $100 to $200 per head. They were looking for someone to blame. Retailers and packers looked like good targets.

So by August 1976, nine suits – alleging collusion and price fixing – had been filed against 23 retail food stores and a dozen meat packers for damages in excess of $1 billion.

The basis for these suits, according to trial lawyers who were advocating them, was the well-publicized Bray case, filed in 1969. Irwin Bray of California and six other cattlemen from California and Colorado were plaintiffs. Joseph L. Alioto, the colorful mayor of San Francisco, and his son were the plaintiffs' attorneys. They charged A&P, Safeway and Kroger with conspiring to fix prices at both the wholesale and retail levels, and subsequently listed other members of the National Association of Food Chains as co-conspirators.

Safeway and Kroger settled out of court for $90,000 (between them) in court costs and attorney fees. But A&P opted to fight it. In 1974, a six-person jury in San Francisco ruled against A&P and awarded $10,904,027 in damages, which was automatically tripled under antitrust law to $32,712,000. A&P later settled for $9,000,000. All of which gave hope to the suffering cattlemen and their opportunist attorneys.

O.J. Barron, Jr., TCFA president from Spur, took an unpopular stand against the suits. "All cattlemen have suffered economically the past two years," he reasoned, "but not just because of retail prices. Our primary problem has been an oversupply of cattle, for which we can blame only ourselves. Total cattle numbers have increased 18 percent in the last 10 years and 33 percent in the last 20 years. . . . Furthermore, food stores this year will spend about $225 million advertising beef compared with $750,000 spent by producers."

Most of the suits were ultimately dropped, primarily because of the Illinois Brick Case in 1977. In this case, the U.S. Supreme Court ruled that sellers were generally precluded from recovering damages if they did not sell products directly to the defendant. Thus, cattlemen could not collect from retailers.

Most cattlemen did not want to sue packers, to whom they did sell directly, although the Meat Price Investigators of Iowa did. They charged in 1976 that IBP, MBPXL (now Excel), Spencer Foods and Flavorland Industries violated antitrust laws by manipulating the ***Yellow Sheet*** (the daily price report published by the ***National Provisioner***).

Spencer and Flavorland agreed to settle and deposited money into an interest-bearing settlement fund, which grew to over $2 million. But IBP and MBPXL chose to fight it.

As late as 1991, the case was settled. But by this time, the Dallas judge could not find proper recipients for damages in the class action suit. So he searched for an organization that represented all cattlemen and awarded $750,000 to the BIC of the National Live Stock and Meat Board. The remainder – over $1 million – went to the trial lawyers involved.

The Beef Research and Information Act, originated by TCFA, was signed by President Gerald Ford on May 28, 1976. Left to right: Charles E. Ball, TCFA; President Ford; O.J. Barron, Jr., TCFA; and C.W. (Bill) McMillan, ANCA.

consumer news reporter from Washington, said, "You (cattle feeders) are victims of the mistrust which has been growing in America. . . . Most consumers are suspicious that the cattle industry is out to rip them off by misrepresenting facts and producing an inferior beef product."

A Checkoff for Beef

Such vocal suspicions and mistrust by consumers underscored the need for a national uniform checkoff for beef promotion and information – an idea originated by TCFA earlier (Chapter 11).

The checkoff idea was discussed at a Managers Roundtable, March 23, 1974 in Amarillo, where President Glenn Deen appointed a committee to study it. The Committee – consisting of O. J. Barron, Jr. as chairman, plus Joe Weatherly of Wheeler, Roy Johnson of Perryton, T. M. Woodley, Jr. of Sabinal and Elton Holland, Jr. of Fort Stockton – met once in April and decided to propose it immediately to ANCA.

The ANCA Executive Committee also liked the idea, so President Gordon Van Vleck appointed a Beef Development Taskforce (BDT) – with Barron as national chairman and Charlie Ball as staff coordinator – to pursue it. The BDT was made up of 15 industry leaders representing eight national organizations.

Over much opposition in Washington and in the country – by consumer groups, American Farm Bureau Federation, National Farmers Union and others – the Beef Research and Information Act passed and was signed into law by President Gerald Ford on May 28, 1976. In the July 1977 referendum, 56.6 percent of the producers voted for it. But a two-thirds majority was required, a requirement put in by Congress at the insistence of the Farm Bureau. So it failed.

A modified bill was passed in 1978, requiring only a simple majority. It, too, failed in a 1980 referendum, when only 34.6 percent of the producers voted for it.

A third bill was written in 1985, the Beef Promotion and Research Act, which was approved by 79 percent of the producers in a 1988 referendum.

Since then, approximately $80 million per year has been collected for promotion and research, half of which is kept by the state beef councils and half of which goes for national programs. A 1991 study by Dr. Ronald Ward, University of Florida, showed that the checkoff has accounted for a 2.24 percent increase in demand (or kept demand from slipping as much as it would have otherwise) and paid the producer a return of $5.70 for each dollar invested.

Consumer Movement Far-Reaching

The consumer movement gained unprecedented momentum during the Carter Administration, 1977-1980. President Jimmy Carter gave it a tremendous boost by appointing Carol Tucker Foreman, president of the Consumer Federation of America, as Assistant Secretary of Agriculture. Hundreds of cattlemen protested with telegrams and telephone calls. TCFA President O. J. Barron, Jr. issued a release, saying, "It is a slap in the face of farmers and ranchers that a consumer activist who is antagonistic to agriculture is being considered for a top position in the USDA." Nevertheless, she and more consumer activists were confirmed.

President Carter, a peanut farmer from Georgia, further riled cattle feeders when he: Stated that "American beef production should move away from grain and toward grass as the main feed ingredient" (1977). . . . Served black bean soup (and no meat) in the White House in observance of National Food Day (1977). . . . Raised the quota on beef imports to let in 200 million more pounds (1978). . . . Vetoed a counter-cyclical meat import bill (1978).

Dietary Goals for the United States

Consumers also had the ear of many congressmen during the 1970s. Sen. George McGovern (D-S.D.), chairman of the Senate Select Committee on Nutrition and Human Needs, conducted hearings in 1976 and on January 14, 1977 released the controversial report, "Dietary Goals for the United States." Among seven "suggestions" in the report was this inflammatory one: "Decrease consumption of meat and increase consumption of poultry and fish."

Cattle feeders had many questions about the market in the 1970s. Answers were frequently provided by economists Ed Uvacek of Texas A&M (left), and Willard F. (Doc) Williams of Texas Tech (right).

Naturally, cattlemen were upset and targeted the report as well as the consumer activists behind it. NCA called the recommendations "arbitrary and with no valid scientific basis." A group of 14 scientists picked by the Council for Agricultural Science and Technology (CAST) concluded that "the goals are not appropriate if applied to the general public." Further, the scientists contended that the goals did not address the two most important aspects of applied nutrition – reducing caloric intake to control obesity and eating a variety of foods to achieve a balanced diet.

So controversial was the nutrition issue, TCFA decided in 1978 to hire a registered dietician – Anne Ilse Anderson – to work with food editors, schools, Extension agents and the public. Her efforts to promote good nutrition, as well as to promote beef, were lauded nationally and in 1984 she became executive director of the Texas Beef Industry Council.

DES Debacle

Although the modern cattle feeding industry has been built on science (Chapter 5), some scientific developments have turned into major issues. Such was the case with diethylstilbestrol (DES), used for humans since 1940 and approved for cattle in 1954. A few dollars worth of DES in the feed or a few cents for an ear implant would increase the rate of gain 15-20 percent and reduce the amount of feed required 10-15 percent. Returns were at least $10 for each $1 spent.

Within a few years, DES became an accepted part of producing beef. That is, until the Food and Drug Administration (FDA) started questioning the possible effect that DES might have on humans eating the beef produced with it.

Cattlemen challenged the FDA, pointing out that no DES was ever found in the muscle of beef – only slight amounts in the liver, which acted as a filter for the animal's system. Further, a person would have to eat 15,000 pounds of liver, cattlemen argued, to get the amount of DES that a woman gets from one birth control pill.

But again, the government sided with consumers, who were questioning the practice. In 1972, the FDA banned the use of DES in the liquid form, while permitting the continued use of supplements and implants, but the issue continued to brew.

On June 29, 1979 – his last day in office – FDA Commissioner Donald Kennedy banned DES as a growth promoter in cattle and sheep. In a release, he contended: DES causes cancer (in rats); residues of DES occur in food (in liver); there is no approved method of detecting residues; and the law does not allow economic benefits to be considered. Manufacture must end by July 13; use must end by July 20.

FDA required cattlemen to remove DES implants from more than 1 million cattle in 1980.

TCFA's Richard McDonald persisted in Austin until ad valorem taxes on livestock were removed, 1980.

Dairy Termination Program of 1986 caused 1.55 million dairy cattle to be slaughtered. TCFA President Mike Bowles (right) and NCA President Don Butler testified against it before the House Agriculture Committee.

TCFA requested an extension for its members (to use up inventory), which was granted – until November 1, 1979. But in early 1980, FDA discovered that a number of operators had continued to implant, "to use up the inventory." And that triggered the "DES Debacle!"

By May 1980, the FDA had cited 254 operators in 23 states for implanting 407,997 cattle after the deadline. That included 39 operators in Texas, 39 in Oklahoma and five in New Mexico. The meat from several hundred cattle was condemned and destroyed; livers and kidneys from thousands of cattle were buried. And the implants had to be removed from the ears of more than a million cattle that had been implanted before the ban.

Cattlemen continued to believe – and do to this day – that there was no risk to human health. Some growth promotents in the production of beef are still approved by FDA, accepted by consumers and used by cattle feeders. In 1989, however, the European Economic Community (EEC) banned beef imports from the U.S., claiming that hormones create a health hazard. The U.S. countered that the EEC hormone ban was in reality an economic trade barrier. And the issue remained unresolved in 1992.

Tax Issues Loom

Another issue that has persisted for at least 20 years for cattle feeders is taxes, especially relating to the prepaying for feed and the cash basis of accounting. During the early days of "The Boom" (the 1960s), it became apparent that "outside capital" was necessary to finance the rapidly growing industry. It also was apparent that "tax advantages," as well as profit potential, was one of the attractions to outside capital.

That soon came under attack by the IRS, however, and has been a recurring issue for TCFA about every two years. The specific issues fell under technical headings like Sub-Chapter S, limited partnerships, farm syndicates, material participation, passive losses, at-risk rules, hedging guidelines, etc.

Economic losses during the wreck of 1973-77 took out most of the outside investors. Nevertheless, the Tax Reform Act(s) of 1976, 1978, 1982, 1986 and 1988 made cattle feeding less attractive to "outsiders" and made the accounting more difficult for all cattle feeders. Thus, taxes have been a major issue for TCFA officers, staff, accountants, attorneys and members for a least two decades.

One of the alltime major legislative victories for TCFA was the 1981 elimination of ad valorem taxes on livestock – an effort that many observers, in and out of the industry, thought was a pipe dream. Ad valorem taxes had been a thorn in the side for Texas feeders – and an expensive one – for years.

TCFA's Dr. Richard McDonald lobbied the Texas legislature relentlessly and convincingly, arguing that the issue was one of fairness: "First, the property tax on livestock is applied very unequally, ranging from a few cents to over $4 per head. Second, some taxing entities tax livestock, some do not, because it's an administrative nightmare to count and assess cattle on the open range. And third, poultry and livestock are the only agricultural products – still in the hands of the producer – that are taxed."

His two-year argument made sense to legislators, who passed the bill in May, and ultimately to the voters, who in November 1981 voted to amend the Texas Constitution to remove ad valorem taxes from livestock.

Antitrust Concerns

One of the most weighty issues ever addressed by TCFA came in 1978, when some corporate feedyard members called attention to the fact that the National Broiler Marketing Association (NBMA), headquartered in Jackson, Mississippi, had been cited for violation of the Sherman Antitrust Act. They requested that TCFA investigate and reaffirm that it was not in violation. The NBMA members had a practice of exchanging supply information, during a conference call each Monday morning, on broilers to be marketed during the coming week. And the U.S. Supreme Court had ruled on June 12, 1978 that integrated broiler producers were not protected from antitrust laws under the Capper-Volstead Act of 1922. The court's rationale was that the corporations were producers, alright, but not as defined by Capper-Volstead "in the production of agricultural products as farmers, planters, ranchmen, dairymen, nut or fruit growers." (Obviously, the Act of 1922 was written before large commercial feedyards came into existence.)

This had implications for corporate cattle feeders. Thus, the question was raised: Is TCFA market information in violation of the Sherman Antitrust Act? Or is TCFA protected by the Capper-Volstead Act?

The stakes were big, since the outcome conceivably could have put TCFA and some large members out of business. All of which prompted TCFA to hire the best legal counsel available – ultimately four law firms in Atlanta, Denver, Houston and Amarillo – who suggested several changes in procedures. For example, TCFA began qualifying members through a membership agreement and limiting market information to members only.

Next, TCFA petitioned the U.S. Department of Justice in November 1978 for a Business Review Letter stating that our new practices were protected under the Capper-Volstead Act. When the letter came on January 31, 1980, the Justice Department declined to say that TCFA was or was not in violation of the Sherman Act, and declined to say whether TCFA was protected by the Capper-Volstead Act.

In the absence of disapproval, TCFA considered it a major victory and has been operating under these procedures ever since.

Dairy Buyouts

Another major issue in the 1980s, ironically, pitted cattlemen against cattlemen – dairy versus beef. The feeling of competition, if not rivalry, had surfaced earlier. In 1933, for example, when the nation found itself into the Great Depression, about 500 members of the Northeastern Panhandle Feeders Association met in Canadian to discuss their plight. Among their recommendations to the government: "Reallocate appropriations so that beef and dairy industries will receive equal benefit."

Then in 1983, the Dairy PIC (Payment In Cash) program was the biggest issue for cattlemen. Milk price supports of $13.10 per cwt. had caused gross overproduction, resulting in Commodity Credit Corporation purchases estimated at $2.367 billion and the need, according to dairymen, to eliminate 1 million dairy cows. So over stiff opposition from TCFA, NCA and other cattle groups, congress passed the PIC, which paid dairymen $10 per cwt. to reduce milk production 5-30 percent.

Dairy Termination Program

But the major battle with dairymen was yet to come – in 1986 with the Dairy Termination Program. While the dairymen accounted for only 18 percent of the country's cattle operations, they had political action committee funds ($2.9 million) that were 13 times greater than what beef cattlemen could collect. Thus, the milkmen seemed to be "smarter lobbyists" than the cowboys.

Cattlemen pulled out the stops but were unable to stop the Dairy Termination Program (DTP), which the USDA announced on March 28 would cause 1.55 million dairy animals to be marketed during the next 18 months at a cost of $1.83 billion to taxpayers. It was a whole-herd buyout program designed to cut the surplus milk supply. Its more immediate effect was to dramatically cut cattle prices.

The week following the announcement, April live cattle futures declined $5.67 per cwt. and June contracts dropped $5.73 per cwt. – the biggest one-week loss ever. Cash prices followed downward $3.50 to $4.50 per cwt. the first week of April – resulting in losses on fed cattle marketed in the Texas Panhandle of $4.5 million. Nationally, losses on all cattle marketed mounted to nearly $25 million.

Edwin L. Cox, Jr., who owned six custom feedyards, bought Swift Independent Packing Co. in 1986 and merged it with Val-Agri.

The Dairy Termination law called for orderly marketing of the surplus dairy cows. But the program announced by USDA was devoid of any semblance of orderly marketing. TCFA and NCA protested to USDA on March 31, but received no consolation. So the TCFA Executive Committee had a conference call, as did NCA, and concluded that the best alternative was legal action against USDA. Five attorneys – three from Amarillo and two from Washington – worked weekends and evenings with staff, documenting the case. Later, in a show of unity, 38 other state cattlemen's organizations joined the suit.

On April 22, 1968, three weeks after the DTP began, cattlemen had their day in court – in Lubbock before Federal Judge Halbert O. Woodward. On April 30, he handed down a ruling in favor of cattlemen and gave the USDA until June 1 to write and implement regulations in compliance with the law. This – plus congressional hearings, trips to Washington and much lobbying by cattlemen – prompted a settlement agreement on May 9, between NCA President Don Butler and Agriculture Secretary Dick Lyng, and cattlemen's six-week quest was resolved. At least, for the time being.

To the astonishment of cattlemen, however, similar legislation was proposed again in 1991. It, too, was defeated by cattlemen. At least, for the time being.

But on October 28, 1991, NCA received a threatening letter from the National Milk Producers Federation deploring cattlemen's opposition to dairy legislation and stating that "we will aggressively pursue our industry's objectives, irrespective of their impact upon NCA . . . (and) inform dairy farmers across the country that NCA's political objective is to bankrupt dairy farmers. Every dairy farmer can then determine if it is any longer in their interest to contribute to the Beef Promotion Board checkoff program."

Packer Control of Inventory

Another issue of continuing concern to many cattle feeders is packer control of inventory. As early as the 1920s and 30s, Swift and some other packers were feeding a goodly percentage of cattle fed in Texas, ostensibly to assure a supply. The Packers and Stockyards Administration (P&SA) started collecting figures on packer feeding in 1955, when packers accounted for 5.1 percent of all fed cattle marketed. (The number fed by packers has increased substantially since then, but the percentage has not changed a great deal, if forward contracted cattle are not included.)

In the early 1970s, P&SA promulgated Regulation 201.70a, which stated, "Packers are not to own or finance custom feedlots; custom feedlots are not to own or finance packers." Thus, Glover Packing Co. of Roswell and Amarillo was prohibited from custom feeding in their recently purchased 40,000-head Diamond A Feedyard at Roswell and, in 1974, sold it back to Robert O. Anderson.

In the late 1970s, IBP bought a nearly new plant in Boise, Idaho from Missouri Beef Packers and entered into a financial arrangement with three large feedyards to share profits. So, following Regulation 201.70a, the P&SA sought a preliminary injunction in a U.S. district court to stop the arrangement, and lost. Consequentially, P&SA dropped that regulation and substituted for it a meek "policy statement" that packer ownership of custom feedlots "presents problems" and encouraged anyone considering such an arrangement to consult P&SA.

This opened the door for more packer control of inventory.

In 1982, when Cargill, Inc. purchased Excel Corporation (Cargill already owned Caprock Industries and its four feedyards), the P&SA investigated but took no action.

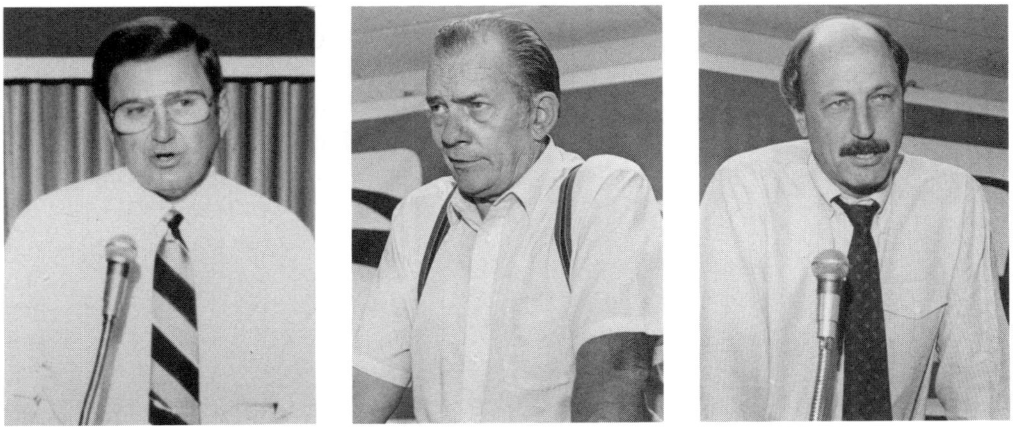

Packer control of inventory became a major concern of cattle feeders. Presidents of the three major packers appeared before TCFA Managers Roundtables in 1988 to explain their position. Left to right: Bob Peterson, IBP; Ken Monfort, ConAgra Red Meat; and Bill Fielding, Excel.

Then in 1986, Edwin L. Cox, Jr. of Dallas (and a few other investors), who owned six custom feedyards, purchased Swift Independent Packing Company and merged it with his Val-Agri packing plant.

Meanwhile, by the late 1980s, some TCFA members were uneasy about the growing control by packers over fed cattle numbers, especially with the increasing use of forward contracting. (Forward contracting is where a packer signs a contract to buy cattle for future delivery, makes a small down payment and normally hedges the cattle.)

In July and August 1988, TCFA invited the presidents of the three major packers to speak on packer control of inventory, at Managers Roundtables in Amarillo. Bob Peterson, president of IBP, said his company had no interest in forward contracting, said it was bad for the industry and urged managers not to participate. "But if forced into feeding and contracting, to meet the competition of our two major competitors, we will become big....and probably purchase feedyards and then feeder supplies." (A month later, Cactus Feeders, Inc. announced that they had entered into a contract with IBP to sell the bulk of Cactus' cattle to IBP on a formula basis.)

Lester Norton, publisher of the ***National Provisioner****, spoke to TCFA directors in 1978, and refuted claims that his* ***Yellow Sheet*** *hurt cattle prices.*

Ken Monfort, president of ConAgra Red Meat Companies, said, "We feed about 20 percent of our kill and have no plans to go past that." Bill Fielding, president of Excel, said, "Excel feeds about 4 percent of its total kill. In addition, we forward contract 40-50 percent at times, and only 20-30 percent at times. And we don't anticipate we'll expand beyond that."

At that time, TCFA started publishing in its weekly newsletter the number of Contract Cattle sold, as reported by managers, but the Market Department says the figures are "not very complete." That same year, 1988, NCA appointed a Concentration/Integration Task Force (TCFA's representative was Leo Vermedahl), which concluded that "free enterprise should be allowed to continue."

Myriad of Major Issues

Among the myriad of other major issues that TCFA addressed through its 25 years: Emergency livestock loans for feeders, from FmHA and SBA (1974-1975). . . . Suit against P&SA, resisting registration and bonding of feedyards (filed by TCFA in 1974, won in 1976). . . . Lifting of Texas state sales tax on major purchases by feedyards (1974). . . . Pressure on FDA to permit continued use of antibiotics in feed (1978). . . . Supported Meat Pricing Task Force, appointed by USDA (Ron Davenport was TCFA representative), seeking answers to "manipulation" and "price fixing" (1979). . . . Interest rates that jumped, jumped, jumped to 21 percent (1979-1980). . . . Grain prices reached new highs in 1980, to $7.35 per cwt. for corn, $6.80 per cwt. for milo.

Still other major issues: An NCA committee to scrutinize futures trading found no evidence of collusion or wrongdoing (1981). . . . Feeders Lean Beef started for TCFA members, to meet consumers new interest in lean (1982). . . . Trucking was partially deregulated for Texas agricultural products (1983 and more in 1987, 1989, and 1991). . . . Amarillo was added as a CME delivery point, Guymon was dropped (1984). . . . IRS declared war on tax shelters, required some feedyards to register (1984). . . . Packer payment for cattle again became an issue, as packers began issuing two-party and three-party checks (1985). . . . Clear title for buyers of Texas agricultural products (1985-87).

And there were more major issues: Texas Beef Council was officially formed in 1986. . . . A Beef Quality Assurance Program was announced by TCFA to assure consumers and others that we produce a safe and wholesome product (1986). . . . Controversy over cattle futures resurfaced; an NCA Task Force negotiated for changes in CME contracts to make them more useful for cattlemen (Zay Gilbreath represented TCFA, 1986). . . . An arbitrary cap on cash accounting – $25 million in gross receipts – was forced on cattlemen by Congress (1987). . . . "Black Monday," October 19, 1987, slammed the stock market and cattle futures; caused the Dow Jones average to drop 500 points in a week and live cattle futures to go down 410 points the following week. . . . Clear title resurfaced as an issue in the Texas Legislature, causing cattlemen and bankers to lock horns (1989). . . . The European Community banned U.S. beef produced with growth promotents, a move motivated by politics, not health concerns (1989). . . . In-bond feeding of Mexican cattle in the U.S. became a reality (1990). . . . The Texas Water Commission ruled that the Edwards Aquifer was "a river," making it subject to state regulations as surface water (1992).

During the past quarter century, as TCFA and the industry resolved a major issue, it appeared that another one always cropped up. Which suggests an on-going role for TCFA in the years ahead.

Environmental activists demonstrated outside the National Cattlemen's Association Convention in Dallas, 1991.

ANIMAL WELFARE VERSUS ANIMAL RIGHTS

Nothing in this century has perplexed cattlemen as has the so-called "animal rights" movement. It's perplexing because cattlemen depend on animals for their livelihood and pride themselves on "animal welfare." But "animal rights?" That implies God-given rights, which most cattlemen believe is reserved for humans.

As a social movement, animal welfare/rights has recurred periodically throughout man's history. Salt's book, ***Animal Rights***, published in 1892, refers to other writings in the U.S. as early as 1725. But the current frenzy began in 1964 with the book ***Animal Machines***, written by Ruth Harrison, an English vegetarian. It was followed in the 1970s and '80s by other books – ***Animal Liberation, Animal Factories, Factory Farming, Modern Meat*** and others – that captured the fantasy of city activists, mostly women.

In 1992, there are about 7,000 "animal protection" groups in the United States, according to the Animal Industry Foundation, ranging from legitimate animal welfare groups to the fringe animal rights "cell." The 400 or so hard-core "animal rights" organizations control a combined budget of $50-75 million a year – more than the combined budgets of all cattlemen's associations in the U.S.

Protesting what they call "cruel confinement" and "unsafe food," the activists use such slogans as "Go Vegetarian," "Meat Is Murder," "Meat Is Dead," "Respect Animals, Don't Eat Them," and "Beef Stinks." They believe that animals are not to be used by humans for any purpose – not for food, clothing, scientific research, education, entertainment or any other use. Some claim that animals have the same rights as humans.

Probably the most radical and certainly the most successful group is People for the Ethical Treatment of Animals (PETA). Founded in 1980, PETA raised more than $9 million from its 350,000 members in 1990. It employs about 90 people and has offices in Washington, D.C. PETA's stated belief is: "Animals do not belong to us to eat, wear, or experiment on."

Other well-known groups: American Society for the Prevention of Cruelty to Animals, Humane Society of the United States, Humane Farming Association, Farm Animal Reform Movement (FARM), Farm Sanctuary, Earth First, Animal Liberation Front and Animal Welfare Institute.

The cattle industry has not stood idly by. They have tried to avoid confrontation, however, recognizing that these groups thrive on publicity which results from confrontation, misinformation and the bizarre.

The cattle industry approach, led by NCA, has been to work through such entities as an Animal Welfare Issues Management Team, an NCA Animal Care Subcommittee, the Farm Animal Welfare Coalition and the Animal Industry Foundation. Some of the strategies include a Strategic Plan on Animal Welfare; legislation for the protection of animal research facilities; videotapes and brochures on "Cattlemen Care;" development of third-party experts who can disseminate facts about the beef industry (scientists, physicians, legislators and other respected professionals); etc.

So far, these strategies have worked well for cattle feeders and the cattle industry.

Amarillo Globe News

First officers of TCFA, 1967 (left to right): Jack Carrothers, president; Bob Carter, vice president; Robert Allen, secretary-treasurer.

Photo by Wyatt McSpadden

Fifteen past presidents gathered in Amarillo, March 1992, to help plan the 25th Anniversary of TCFA. Left to right, front row: Joe Hathoot, Wes Bonner, Dick Waterfield, Charlie Koontz. Second row: Leo Vermedahl, Zay Gilbreath, Bob Bliss. Third row: Clark Willingham, Leon Miller, Tom Herrick, Bob Carter. Fourth row: Jay O'Brien, Mike Bowles, Glenn Deen, Bob Josserand.

Chapter 10

TCFA LEADERS

Men who possessed character, commitment, boldness, fairness and vision.

During its 25-year history, TCFA has been blessed with effective leaders, men who possessed character, commitment, boldness, fairness and vision – characteristics that spelled "leadership" and led to Association growth. Almost without exception, they have been "conservative risk takers" who understood the proper relationship between the volunteer leader and staff; they delegated; they inspired; and they fostered progress.

TCFA has had 18 presidents. The first seven served two years each; then the precedent was established for a one-year term.

Jack Carrothers

JACK CARROTHERS
1967-1969

When Jack Carrothers of Friona was elected the first president of TCFA on May 18, 1967, he took the helm of an untried ship and set out on unchartered water. He proved to be a man with uncommon conviction, vision, determination and leadership abilities. When Jack spoke, people listened.

During those first two years, Carrothers took the lead in protecting the fledgling cattle feeding industry from excessive government intervention, especially in areas like water quality control, land use planning and other environmental issues.

Under his administration, a market information department was established, the first research projects were sponsored, the current health insurance program was initiated and an aggressive national beef promotion program was launched.

Early into his term, TCFA was denied membership in the American National Cattlemen's Association, so he led the TCFA delegation in storming out of the ANCA board meeting in Oklahoma City. A few months later, however, TCFA was admitted.

Carrothers was a native of Colorado and for several years was a cattle buyer for Armour & Co., after graduation from Oregon State University in 1950. He moved to Texas in 1961 (from Colorado where he was in the cattle feeding business since 1954), helped build Friona Feedyard, and became president of Friona Industries, Inc. (a publicly-held company). He died in 1977.

R.M. (BOB) CARTER
1970-1971

R. M. Carter

"Growth" was the watchword during the term of Bob Carter of Plainview. New feedyards sprang up across the state as the number of fed cattle marketed increased from 2.7 million head in 1969 to 3.6 million in 1971.

Carter now looks back on "a career of beef promotion." He became TCFA's representative on the National Live Stock and Meat Board in 1972 and was elected chairman for 1982-83. During his term as TCFA president, the first "Beefiesta" was conducted in Dallas where more than 60 home economists and food editors learned firsthand about Texas Grain-Fed Beef. As a member of the Cattlemen's Beef Promotion and Research Board, which administers the checkoff, he serves on its advertising committee and executive committee. Also, in 1974-75, he was Regional Vice President of ANCA.

Even as early as his presidency, TCFA began work to protect feeders from slow paying packers who were "floating" drafts. Also, stepped-up research on shipping fever was sponsored at Texas Tech, Texas A&M University and Oklahoma State University.

A native of Plainview, Bob attended Texas Tech and graduated from the University of Oklahoma. He and a partner opened the Carter-Kirchhoff Feedyard in 1955, which in 1986 evolved into C Bar Feedyard in Plainview.

TOM HERRICK
1972-1973

Tom Herrick

Boycotts, price freezes and the energy crisis made headlines during Tom Herrick's term. There also was a ban on DES in feed, the "worst winter in history" and scabies quarantines.

But it wasn't all negative during his term as president of TCFA. Fed cattle prices hit a new high of $58 in August 1973; Texas became the No. 1 cattle feeding state; and an emergency fund was raised "to tell the cattlemen's story." Herrick exhibited charisma and decisiveness that led people to follow him.

As cattle prices began to edge upward in 1972 and early 1973, consumers became discontent and began beef boycotts, prompting President Richard Nixon to slap a price freeze on beef. In April 1973, Herrick called an emergency meeting of the TCFA Executive Committee and proposed that TCFA members contribute $1 per head to an emergency fund. Within three weeks, feeders had contributed over $519,000. These funds were shared by TCFA with ANCA and the Meat Board, and it turned around all three organizations. Those organizations, all struggling financially at the time, have been growing ever since.

Herrick, a former Colorado order buyer, moved to Amarillo in 1959 and built Randall County Feedyard (with three partners) in 1966-67.

Glenn Deen

GLENN DEEN
1974-1975

Feedyards faced their second and third straight years of financial problems during Glenn Deen's term as president of TCFA. He took steps to help feeders minimize their losses.

In April 1974, TCFA made several recommendations to ANCA, including: a national checkoff for beef; an expanded economic analysis staff; beef grade changes; increased Washington, D.C. manpower; and monitoring the import-export situation. All were accepted and accomplished by ANCA.

Deen was a master at cooling a hot situation or flared tempers in a meeting. And there were plenty of each during "The Wreck." He appointed a special study committee, chaired by O. J. Barron, Jr., to examine new ways to help the cattle industry through the doldrums. This committee proposed a plan which evolved into the national $1-per-head checkoff.

During his term, TCFA also expanded market news and membership services and made them available to feedyards in Oklahoma and New Mexico. In 1977-79, he was Regional Vice President of ANCA and, in 1979-80, President of Cattle-Fax.

Deen, a native of New Mexico and a 1955 graduate of New Mexico State University, managed Chow-Tex Feedyard at Lazbuddie (1965-66) and moved to Dumas in 1967 to become part owner and manager of Texas Beef Producers.

O.J. Barron, Jr.

O. J. BARRON, JR.
1976-1977

Several new programs for TCFA members were initiated while O. J. (Jim) Barron, Jr. was president. He made 21 trips toWashington, D.C., plus an equal number to other states, and became known as "the TCFA traveling president."

Barron helped in the passage of a Federal prompt payment law and in finalizing the long-pending suit against registration and bonding of feedyards. The U.S. Meat Export Federation was organized early in his term (TCFA became a charter member); and the Beef Research and Information Act, which ultimately led to the $1-per-head checkoff, was signed into law by President Gerald Ford. Also, through the influence of TCFA, merger plans for the American National Cattlemen's Association and the National Livestock Feeders Association were started.

Barron, a native of El Paso, attended the Texas College of Mines and the University of Arizona. Following World War II, he became owner of Spur Headquarters Ranch and feedyard at Spur, Texas.

Leon Miller

LEON MILLER
1978-1979

The first year that Leon Miller of Eagle Pass was president, 1978, started off with a roar – the roar of tractors driven to Washington by 5,000 members of the American Agriculture Movement. The following two years were like a roller coaster. Cattle prices soared from $43 per cwt. in January to $62 in June. Then President Jimmy Carter opened the gates to imports, triggering a $10 break in prices. All year long, TCFA, NCA and others worked for a counter-cyclical import bill, only to have it vetoed by President Carter.

During Miller's second year as president, President Carter signed a modified import bill; FDA Commissioner Donald Kennedy banned DES as his last act in office; beef boycotts fizzled; and Choice steers closed at a then-record $70 per cwt. So Leon left office when the "roller coaster" was at a high.

Miller was born in Texas, grew up in Missouri, attended the University of Missouri and Colorado State University, built and managed a new feedyard in Colorado, managed a "dehy" and supplement operation in California before moving back to Texas. He and a partner started Alta Verde Industries at Eagle Pass in 1966, then he bought out the partner in 1970.

R. L. (Bob) Bliss

R. L. (Bob) BLISS
1980-1981

Bob Bliss' term as president of TCFA was filled with unusual challenges. In 1980, grain prices hit new highs – $7.35 per cwt. for corn, $6.80 for milo. And the prime interest rate reached 21 percent. Then came the DES debacle, when FDA cited 254 operators for illegally implanting 407,997 head of cattle with DES. The Beeferendum was turned down, for the second time in three years. And in 1981, the kangaroo meat scandal focused attention on meat imports. (Kangaroo meat actually came in from Australia, labeled as beef.)

But in his second year, events became more positive. Ad valorem taxes were eliminated on livestock in Texas. And TCFA reduced dues 20 percent to 40 cents per head.

Bliss, a native of Iowa, came to Texas in 1962, first to work for King Feed Yard at Muleshoe; then as builder-manager-part owner of Swisher County Cattle Co. in 1967. Later he was associated with Mesa Agro, Friona Industries and Caprock Industries. He was a director of the Meat Board (1980-81) and president of Cattle-Fax (1984-85). Known as a teacher, he trained at least eight young men who became feedyard managers.

Gene Newman

GENE NEWMAN
1982

Gene Newman of San Angelo served only one year as president and established that precedent.

Among the highlights of his term: (1) TCFA initiated volume grain shipments and reduced rail rates for Midwest corn to Texas feedyards. (2) Feeders Lean Beef was introduced, a marketing association through which 20 TCFA feedyards sold short-fed beef directly to retailers. (3) TCFA adopted a long-range plan, which has been updated each year since and serves as a "road map" to keep directors and members headed in the same positive direction year after year.

Newman, a graduate of Sul Ross State University at Alpine, was a rodeo performer, a cutting horse trainer, a pilot, an art collector, a tough competitor and a "gung ho" civic leader. While president of TCFA, he also was on the executive committee of NCA, both organizations which he supported with enthusiasm and fervor, until his death in 1985.

Charles E. Koontz

CHARLES E. KOONTZ
1983

Like several presidents before him, Charlie Koontz of Olton felt the brunt of a swift market drop during his term – from $72 per cwt. in April 1983 to $58 in September. Part of the blame he placed on the Dairy PIC (Payment In Cash), "which bailed out dairymen but left cattlemen holding the bag." Then came the Grain PIK (Payment In Kind), which escalated feed costs.

TCFA established a private market wire during his tenure; and sponsored a unique Think Tank in Dallas, where scientists zeroed in on the problem of warmed over flavor in beef.

Koontz, a native of Arizona and a graduate of California Polytechnic State University, operated feedyards in California and Hawaii before moving to New Mexico in 1966 to manage Janss Feedyard at Roswell. Then he moved to Texas in 1969 as builder-manager-part owner of Olton Feedyard at Olton. With a deep voice and a quick laugh, Koontz handled controversy in meetings well. He was the TCFA representative on the Texas Animal Health Commission, 1973-1987, and on the Cattlemen's Beef Board, 1986-1991.

Dick Waterfield

DICK WATERFIELD
1984

The year 1984 was a milestone for Dick Waterfield and for TCFA – he served as president and Texas fed 5,000,000 cattle, the biggest feeding year for any state in history. Including Oklahoma and Texas, a record 6,000,000 head were fed in the TCFA area.

The first serious threat by Congress on prepaid feed and cash basis of accounting was during his term. But cash basis was saved, and TCFA directors voted to make increased legislative activity the No. 1 objective.

As a result of all this, Waterfield became interested in politics, was elected Texas state representative in 1986 and ran (unsuccessfully) for U.S. Congress in 1990.

Dick was a disciplined jogger – three to five miles before breakfast – and a dedicated Sunday school teacher and lay minister. A native of Canadian, Texas, he majored in business at Oklahoma State University. His feeding experience started in 1966 with Beef Cattle Co. at Wheeler, then to Canadian Feedyard in 1972 as manager and part owner, then to Follett Feeders in 1984 as manager.

Robert D. Josserand

ROBERT D. JOSSERAND
1985

Commanding, imposing, demanding, bold, effective are a few of the adjectives used to describe the leadership of Bob Josserand of Hereford. And that kind of leadership paid off while he was president of TCFA in 1985. Among other things, the Beef Promotion and Research Act was adopted, authorizing the $1-per-head checkoff and a delayed referendum on whether to continue it. Also, clear title legislation became law, giving buyers of ag products some protection against double payment.

While a director of the Meat Board (1978-1983), Bob gave a "Oneness of Effort" speech in 1981 that leaders still talk about. His efforts are credited with increased revenue for the BIC – from $2.8 million in 1981 to $13.7 million in 1985. He became president of NCA in 1989 and has served on the Cattlemen's Beef Board (1986-92).

A native of Kansas, Josserand graduated from Colorado State University in 1953, managed a feedyard in Kansas, moved to Texas in 1971, became president of Pro Cattle Co. (1973-1979) and AZL Cattle Group (1980-1983), bought the company in 1983 and changed the name to AzTx Cattle Company. AzTx continued to grow under his leadership, to become by 1988 the seventh largest cattle feeding operation in the country.

Mike Bowles

MIKE BOWLES
1986

The Dairy Termination Program was a dominant issue during the 1986 term of Mike Bowles, who fought it from his home office in Hereford, in the Halls of Congress and in the courts. When the USDA announced the DTP, it sent shock waves through the beef industry, driving cash prices down $4 to $5 in less than a week. TCFA and other cattlemen's organizations acted swiftly and filed a suit against USDA for not complying with the law. After cattlemen won, Bowles said, "This proved one thing to cattlemen – we need to get more involved in the political process and we need more money for our Beef-PAC."

Among other significant events during his tenure: The $1-per-head checkoff began on October 1, 1986, although the producer referendum making it permanent was yet to come in 1988. The Tax Reform Act of 1986 authorized some of the most sweeping tax law changes in history. TCFA announced a Beef Quality Assurance Program, an activity he had long practiced and advocated. He helped organize the NCA Beef Safety Assurance Task Force and served as chairman for six years.

Bowles, a native of California and a graduate of California Polytechnic State University in 1963, came to Texas in 1967 as a feed supplement salesman. His feedyard experience followed as manager of Three C Cattle Co. at Lubbock, 1972-1977; Pre-Feeders at Summerfield, 1978-1989; and Swisher County Cattle Co., 1990-1991.

Zay W. Gilbreath

ZAY W. GILBREATH
1987

Zay Gilbreath, TCFA president in 1987, may have been the only president to that time who could boast of "12 consecutive months of cattle feeding profits." He also can boast about the industry collecting for the first time $73 million for beef promotion and research (under the new $1-per-head checkoff).

During his tenure, TCFA instituted M-Net, a first-of-its-kind computer-satellite market information system. Also, an expanded safety program for feedyards with a full-time safety director. Beef-PAC set new records and made political contributions totalling $73,500.

But not all the clouds had silver linings during his term. Controversy over cattle futures surfaced, and Zay was appointed to an NCA Futures Task Force to investigate. An arbitrary cap on cash accounting ($25 million in gross receipts) was forced on cattlemen by Congress. And "Black Monday," October 19, 1987, slammed the stock market and cattle futures. He served as a director of the Meat Board, 1986-89.

Gilbreath, a native Texan who graduated from Texas A&M and Kansas State University in agricultural economics, joined Cargill in 1963. Since 1976, he has been with Caprock Industries, Cargill's feeding subsidiary, where he was promoted to President in 1981.

Clark S. Willingham

CLARK S. WILLINGHAM
1988

When Clark Willingham received the telephone call inviting him to become an officer of TCFA, he could hardly believe it. After all, he was a Dallas attorney. But he was a different kind of attorney – an agricultural tax attorney, with part-ownership of Hitch feedyards in Oklahoma – and cattle feeders had major concerns about taxes. Clark also had a special interest in legislation – had a part-time office in Washington and knew many congressmen on a first-name basis.

Clark didn't solve all the tax problems, but a number of positive things happened during 1988, his year as president. Cattle numbers dropped to 99.6 million, the first time under 100 million since 1961. Fed cattle prices averaged above $71 (at $71.08) for the first year ever. The beef referendum, on the $1-per-head checkoff, which had failed twice before, passed with a 79 percent majority. Packer control of inventory became an issue and TCFA hosted the presidents of the "Big 3" packers to discuss the issue with managers.

Willingham, a native of Dallas, received his BBA degree in accounting from Texas Tech in 1967 and a JD degree in 1971 from Southern Methodist University. He has been a regional vice president of NCA (1991-92), Chairman of the Texas BIC, a director of the Meat Board and the U.S. Meat Export Federation.

Wesley B. Bonner, DVM

WESLEY B. BONNER, DVM
1989

When food safety began to surface as a national concern, TCFA was fortunate to have a veterinarian with experience in human clinical research standing in the wings. Dr. Wes Bonner, TCFA president in 1989, had a lot of credibility in the cattle industry and with the public. And he made food safety a priority. Recognizing his expertise, the NCA, the U.S. Meat Export Federation and the USDA have called on him frequently – as a consultant and as a spokesman, in this country and abroad.

During his term, the European Community banned U.S. beef produced with growth promotants, a move that he says was motivated by politics, not health concerns. But that same year, Mexico opened its borders for more cattle trade. Fed cattle prices rose above $80 for the first time ever. And TCFA embarked on a plan to tell the world why the TCFA area is "Cattle Feeding Country."

Bonner, a native of San Angelo and a graduate of Texas A&M University in 1969, got his feedyard training at Veribest Cattle Feeders under the tutelage of an earlier TCFA president, Gene Newman. In 1977, Bonner and five other local stockmen bought the feedyard.

Leo D. Vermedahl, Ph. D.

LEO D. VERMEDAHL, Ph. D.
1990

Like many entrepreneurial feedyard managers before him, Leo Vermedahl, TCFA president in 1990, was attracted to Texas by the opportunities in the budding new industry. After earning a Ph.D. at the University of Minnesota in 1968, he became a consulting nutritionist for Allied Mills. Then he heard about a new Feedyard Management Program at Texas A&M. So in 1971, Leo and family headed for Texas.

An internship found him working with Bob Bliss and Tom Herrick – both past TCFA presidents. After graduation in 1972, he stayed on with them as assistant manager of Swisher County Cattle Co. at Tulia. Since then, he has been manager and/or part owner of the following: Coronado Feeders at Dalhart (1975-1979), Granada Cattle Feeding Services (1980-82), TBP Group Feedlots at Dumas (1983-1985), Seven X Feedyard at Summerfield (1989-1991), and in 1992 joined Tejas Feedyards at Pampa.

NCA, as well as TCFA, has called on his expertise frequently. He has served as chairman of the NCA Feeder Council, on the NCA Beef Industry Concentration/ Integration Task Force and the NCA Executive Committee.

One thing Vermedahl jokingly takes credit for, while president in 1990: Fed cattle prices hit an all-time high of $82.50 per cwt. and traded at an average of $78.43 for the year.

John Jay O'Brien

JOHN JAY O'BRIEN
1991

Student, intellectual, historian, individualist, entrepreneur. These are but a few of the monikers that peers have used in attempting to describe Jay O'Brien, the 1991 president of TCFA. None of these quite describe him. For he is an unusual, if not an unlikely, cattleman. He doesn't look, act, dress or talk like a typical cattleman. But he is a successful rancher, stocker operator, cattle feeder and feedyard owner – all on a sizeable scale.

Jay describes himself as an environmentalist and a cattleman. He defends cattlemen's environmental record and advocates working with other environmental groups. He has served as vice chairman of the NCA Private Lands and Environmental Management Committee, on several other NCA committees, as well as the NCA board. Having a keen interest in education, he championed the TCFA Education Foundation, aimed at attracting bright young people into the cattle feeding industry.

O'Brien, a native of Amarillo, was a 1967 graduate of Yale University, with a major in American Studies (history). Then he became a field editor of ***CALF*** News, a position calculated to give him "the opportunity to travel, meet, interview and study the best cattlemen in the business."

Joe Hathoot

JOE HATHOOT
1992

The 1992 president of TCFA, Joe Hathoot of Canadian, brought a unique perspective to the Association – that of marketing. After graduating from Oklahoma State University in 1957, he ranched a few years in Oklahoma, then joined with Wayne Owen and managed a livestock auction in Texarkana, Texas. (They later built a feedyard at Morton, Texas, and Owen became a director of TCFA.)

During the next 20 years, Joe became a director of the Oklahoma Cattlemen's Association, a director of the National Livestock Marketing Association and President of the Texas Livestock Marketing Association. Since 1987, he has been the TCFA representative on the Texas Animal Health Commission.

Early in his term, the TCFA Environmental Quality Assurance Notebook was published, which he called "another step in the TCFA environmental soundness program."

"Even though TCFA has had a tremendous 25 years, and has benefitted its members in many ways, the challenges for cattle feeders will be even greater in the years ahead," concludes Hathoot. "The need for group action will be even greater. And the opportunities for TCFA will be even greater."

Photo by Wyatt McSpadden

Fifteen past presidents of TCFA, representing over 500 years experience in cattle feeding, met in March 1992 to discuss the future of cattle feeding.

First four presidents of TCFA appeared on an outlook panel at the 1972 TCFA Convention. Left to right: Glenn Deen, 1974-75; Bob Carter, 1970-71; Jack Carrothers, 1967-69; Tom Herrick, 1972-73.

In 1982, TCFA members headed up three national organizations (left to right): W. J. (Dub) Waldrip, president of NCA: R. M. (Bob) Carter, chairman of Meat Board; George Feaster, chairman of U.S. Meat Export Federation.

Photo by Wyatt McSpadden

Three chief executives of TCFA (left to right): Lloyd Bergsma, 1967-72; Charles E. Ball, 1972-88; and Richard McDonald, 1988-).

Chapter 11

THE PROFESSIONALS

TCFA executives have preached, practiced and promoted professionalism.

According to conventional wisdom in association management, "Associations work best when committees recommend, boards decide and staff executes." That's the way TCFA has worked through the years – a "separation of power" between volunteer leaders and the staff leader – and that's one key to TCFA's success.

TCFA has had three chief executives during its history: Lloyd Bergsma (1967-1972), Charles E. Ball (1972-1988) and Richard McDonald (1988-present).

THREE EXECUTIVES – THREE APPROACHES

LLOYD BERGSMA: *With vision, he initiated "services that members wanted and were willing to pay for." Such as market information, group insurance, beef promotion, cowboy workshops and managers roundtables. Several of his programs are still in existence – 25 years later.*

CHARLIE BALL: *Often observing that "Associations run America," he avidly and steadfastly promoted leadership, professionalism, excellence and bold ideas in association management. Said he, "Only those willing to go too far will know how far they can go."*

RICHARD MCDONALD: *With a Ph.D. education, an analytical mind and a penchant for thoroughness, he excelled early in the regulatory and legislative areas. "He does his homework, he's credible and reliable," said legislators. The result was a number of legislative and regulatory victories.*

Lloyd Bergsma, a former Extension marketing specialist, helped organize TCFA and put in place management practices that have remained 25 years.

Lloyd Bergsma

Lloyd Bergsma set a pattern of professionalism that has continued for 25 years. He had a good background, a strong personality and was an efficient organizer. Reared on a farm in western Kansas, he attended Kansas State University and Colorado State University, where he graduated in 1958 with a degree in agricultural economics. Between the two university stints, he managed the family feedyard for 10 years. Following college, he worked for Doane's Agricultural Service in farm management and later as Livestock Marketing Specialist at Texas A&M University.

When the Texas & Southwestern Cattle Raisers Association started its Cattle Feeders Division, in 1961, they hired Bergsma as the new Director. In early 1967, he resigned from TSCRA, helped organize the new TCFA (as a consultant), then became its first executive officer.

Among his first duties were to draw up a proposed budget and write new bylaws. The annual budget he submitted was for $36,675 in expenses, including a modest $15,000 for his own salary. But Jack Carrothers, the new TCFA president and cattle trader that he was, slashed $3,000 from the proposed salary, handed the budget back to Lloyd, and they had a deal. He made out better with the proposed bylaws. They were checked by the Association's attorney, James W. Witherspoon of Hereford, who commended Lloyd for his work and the new board accepted them.

Even with the reduced salary, Lloyd went some months without a paycheck. "I wasn't worried," he says now. "I never had a doubt that this Association would be successful. I knew the industry was destined to be in the Southwest and I knew TCFA was destined to serve it."

To get the Association going, he tells about 10 cattle feeders who agreed to put up $500 each in advance dues. "Two of these never came through," he says, "but the other eight made up the difference and the principal was all paid back within two years." The initial dues were $25 per regular member, plus 10 cents per head of cattle marketed.

Highlights of the early years for Bergsma and the new association occurred rapidly. He lists five:

■ "Having a live, viable organization that could serve the rapidly growing industry was the primary one," he says as he looks back. "That required money. But within two years, revenues had tripled to over $107,000 and several needed services were in place. (Also, his salary was adjusted to a respectable level.)

■ "Timely market information, with a wire service, was necessary for our success. This was difficult at first, because Cattle-Fax was a going operation before us, and they

Bergsma (right) initiated a beef promotion in 1969 with TAP. Shown with Agriculture Commissioner John C. White (left) and Gov. Preston Smith.

Bergsma (left) makes presentation to TV star James Brolin, a beef promoter, at the 1970 Beefiesta in Dallas.

wanted to take over all market information. But as soon as we could afford a fulltime market reporter, we became the preferred source for market information (for our members) and it continued that way.

■ "From day one, environmental issues were a concern of feedyards and, thus, an important issue for TCFA. Some fish had died in Buffalo Lake, on Tierra Blanca Creek below Hereford, and a young newspaper reporter pointed the finger at our feedyards. But by working with the newspaper publisher and the director of the new Texas Water Quality Board, we pretty well resolved that.

"The Texas Air Control Board also was new and looking for something to do. They were not a big problem, however, because they had no way of measuring odors.

■ "Packer payment was an early concern, also. We had a lot of packers, many of which were small and not very strong financially. Typically, they took seven to 10 days to pay and some cattle never were paid for. We worked with the Packers and Stockyards Administration, but they couldn't touch a packer until 48 hours after the cattle were picked up and by then the meat sometimes was gone. That was before the Association got prompt payment legislation (1976), of course.

■ "Research also was important to our members. So we kept a pen of cattle on test at the PanTech Research Center, east of Amarillo. Nearly every pen made a profit the first four years," he concluded with a grin, "and our members liked that kind of research."

To what did Bergsma attribute his success in getting the Association off on the right foot? "I tried to see the big picture and had to tip toe a lot," he answers. "Members say they want a strong executive but most members feel they are the boss.

"I was fortunate to have strong, dedicated leaders," he adds. "As an indication of their dedication, our officers started out paying their own expenses, and they still do in most cases.

"Then we tried to be responsive to member needs and supply the services they wanted and were willing to pay for," he continues. "Among the services we started were market information, group health and life insurance, beef promotion, an Interfirm Comparison Program (now known as INDEX), Cowboy Workshops and Managers Roundtables, to name a few."

A tribute to Bergsma's foresight is that several of his programs and services are still in existence – 20 to 25 years after they were started.

Photo by Wyatt McSpadden

Charles E. Ball, a farm magazine editor for 25 years, brought a communications background to TCFA.

Charles E. Ball

Charles E. Ball, CAE became the second Executive Vice President of TCFA, 1972-88.

A native of Lamar County, Texas, where he won many honors in 4-H Club work, Ball graduated from Texas A&M in 1947, after a three-year interruption in the army during World War II. Then he received an M.S. degree in agricultural engineering from Iowa State University in 1948. Following graduation, he joined the staff of ***Farm & Ranch-Southern Agriculturist*** magazine in Nashville, Tennessee, until 1952. For the next 20 years, he was in Dallas as Southwest Editor of ***Farm Journal***, specializing in cotton and cattle. He also has written for the ***Saturday Evening Post*** and ***Reader's Digest***, and a book on horsemanship, ***Saddle Up***.

At ***Farm Journal***, Ball traveled the U.S. and received a number of awards for his articles, including those about the cattle feeding boom on the Southern Plains. That attracted the attention of TCFA officers and led to his move to TCFA. He appeared on the program of the first TCFA convention (1967 in Amarillo) and has attended every convention since.

While Bergsma's strong suit was in marketing, Ball's initial strength was in communications, public relations and promotion. He revised and renamed the TCFA ***Newsletter***, which soon became the most popular newsletter in cattle circles. Next, he revamped the ***Cattle Feeders Annual***, making it into a combination magazine and reference book.

Ball's expertise in communications began to pay off in 1973, during the beef boycotts and government price freeze on beef. On Saturday afternoon, April 7, 1973, President Tom Herrick called an emergency meeting of the TCFA Executive Committee to discuss what might be done to offset the boycotters and to sooth enraged cattle feeders. The Executive Committee took about one hour to decide that they would ask members for $1 per head to "tell the cattlemen's story," and asked Ball to draft a letter for Herrick's signature. Then they left!

Sunday morning, Ball called in a couple of staff members and they mailed mimeographed letters to over 2,000 members. Three weeks later, they had $519,350 cash on hand! This money was shared with the Meat Board and ANCA, and has been attributed with turning around all three organizations, including TCFA. (See Chapter 7.)

Ten months later, in early 1974, cattle feeders were well into "The Wreck." Losing $100 to $200 per head, they were desperate. The timing seemed right to Ball to propose a national legislated checkoff for beef promotion and research – an idea that he borrowed from the cotton industry. Cotton had a successful $1-per-bale checkoff, which he had written about many times (as an editor of ***Farm Journal***) and helped promote. "I was convinced the idea would work even better for beef than for cotton," Ball said many times, "because (a) cattle is a national commodity, while cotton is regional and smaller; and (b) each beef animal sells an average of 2.5 times, while a bale of cotton usually sells only once." So the idea of a checkoff was proposed at a Managers Roundtable, March 23, 1974.

Well known in Washington, Ball is shown (top) with Secretary of Agriculture Earl Butz, 1974; (left) with Candidate for Vice President Dan Quayle, 1988; and (right) with Sen. Lloyd Bentsen of Texas, 1978.

The idea moved rapidly – from TCFA, to ANCA, to the National Beef Development Taskforce, which Ball served as national staff coordinator. Fourteen years and three beef referendums later, the $1-per-head checkoff became reality.

Meanwhile, back at the office, Ball had put together a highly respected staff of professionals and implemented new programs that members needed and supported. "We must be creative and bold," he reminded his staff weekly. "But remember that we are here for one reason – to serve our members." He quoted maxims and mottos frequently and stuck them on office walls. Maxims like: "We strive for excellence." "Think big." "Bold ideas." "The difficult we do immediately, the impossible takes us a while." "Only those willing to go too far will know how far they can go." And more.

TCFA grew rapidly through the years – in services, in stature and in revenues. Nearly every year was a record year in revenues. This led to a 20 percent reduction in dues in 1980 – from 50 cents to 40 cents per head – unheard of in association circles.

Ball occasionally uses "CAE" following his name. That's for Certified Association Executive, a recognition by the American Society of Association Executives (ASAE) given each year to about 1 percent of the top executives, based on experience, months of study, dozens of books read and a day-long written test. In 1982, when he received the CAE distinction, he had the highest score of any candidate in the United States.

Ball received a number of awards during his career, including: "Outstanding Farm Magazine Writer," 1959, by American Seed Trade Association; "Outstanding Agriculturist," 1971, by Texas Tech University; "Man of the Year in Texas Agriculture," 1983, by ***Progressive Farmer***; and "Twenty-Five Who Made A Difference," 1989, by ***Beef*** magazine.

Photo by Wyatt McSpadden

Richard McDonald, Ph.D., a former Extension beef cattle specialist, brought a technical background to TCFA.

Richard McDonald

Richard McDonald, Ph.D. became the third Executive Vice President of TCFA in 1988. A native of San Benito, Texas, where he won many honors in FFA and 4-H, including Chairman of the State 4-H Council, he graduated from Texas A&M University in 1967 with a B.S. in animal science. Then he proceeded to Louisiana State University where he received an M.S. and a Ph.D. in 1972 in animal breeding. For the next two years, he was Area Livestock Specialist for the Texas Agricultural Extension Service, before joining TCFA in 1974 as Executive Assistant and later becoming Executive Director.

While Bergsma's basic strength was in marketing and Ball's strength was in communications, McDonald's initial strength was in the technical area – research and regulations. This fit well with a growing research program – TCFA has sponsored over 60 research projects through the years. Likewise, with TCFA's ever-concern about the environment and with 15 state and federal regulatory agencies "looking over" feedyards, the members felt more secure with a Ph.D. representing them – a technical person who could speak the language of the regulators.

With a keen mind and the technical training, McDonald became known as a person who did his homework, stood his ground and shot straight. The regulatory people liked this. He got to know them on a first-name basis and on more than one occasion they invited him to help write regulations for feedyards. (Most regulatory people were strangers to commercial feedyards.) Feedyard managers also called on him to help interpret regulations and for help when in trouble with a regulatory agency. Our plan and request was this: If a regulatory agency has a problem with a member feedyard, the agency was to call TCFA first. And it worked most of the time.

While still Executive Director of TCFA (the No. 2 staff position), McDonald also was assigned lead responsibility for Texas state legislative affairs and backup responsibility for national legislative affairs. Again, his thorough, analytical, calculating approach paid off. Typical comments from legislators were: "Richard's style of lobbying is unique – he presents facts, no smoke, no pressure." "He always is welcome, because he doesn't ask for time unless he has significant new information." "He's credible and reliable – we need more lobbyists like him."

This style and this reputation didn't go unnoticed by other lobbyists. Since 1984, he has served as chairman of the Texas Agricultural Council (TAC), comprised of 45 agricultural organizations that represent essentially every general farm organization and commodity organization in the state. When the legislature is in session, the TAC meets frequently (and occasionally between sessions) to discuss upcoming legislation and try to arrive at "a unified front for agriculture." When that is impossible, they frequently form formidable coalitions.

Probably the most impressive legislative victory for McDonald and for TCFA came in 1981 when he organized the coalition, planned the strategy and led the fight to eliminate ad valorem taxes on livestock in Texas. It was a two-year effort, including a constitutional amendment by voters, that saves cattle feeders enough money every year to more than pay their TCFA dues.

Other legislative victories, in which he led or played a major role, include:

■ Tax relief for farmers-ranchers, with land to be taxed on agricultural value rather than market value (1977).

■ Trucking partially deregulated for Texas agricultural products (1983 and more in 1987, 1989 and 1991).

■ Clear title legislation, which prevents buyers of ag products from having to pay twice (1985).

Clear title legislation has been an on-going battle with Texas bankers for at least eight years – bankers wanting to switch to a central notification system, TCFA wanting to retain our hard-won direct notice system. This strained relations at times, but the bankers respected Richard's integrity. For example, McDonald has been a perennial

Photo by Jim Peck

Richard McDonald (right) greets Texas Commissioner of Agriculture Rick Perry during a 1992 visit to TCFA.

McDonald, well known in Austin, shows TCFA's appreciation to Rep. Bill Clayton, Speaker of the Texas House, 1975-83.

speaker at the Bankers Agricultural Credit Conference in Lubbock the past 10 years. And in spite of their differences over clear title, the bankers have continued to invite him to speak.

Back in the office, McDonald has expanded several member services and been instrumental in starting new ones. Such as:

- Three-year plan to promote the TCFA area as "Cattle Feeding Country" (1989).
- Beef Quality Assurance Notebook to assist feedyards in producing a safe, wholesome product (1990).
- Business Information Service that allows members to report bad business deals with suppliers (1989).
- Leadership Training Program for young cattle feeders (1989).
- Junior Fed Beef Challenge (1990), a carcass show for juniors patterned after the show for feedyards, started in 1974.
- Statement of Fundamental Principles, describing the basic operating philosophies of TCFA Feedyard Members (1989).
- TCFA Education Foundation, a scholarship program to attract outstanding young people into the industry (1991).

Photo by Jim Peck

Feeder cattle, good weather, credit and professional management make this "Cattle Feeding Country." Tascosa Feedyard, 1992.

Chapter 12

CATTLE FEEDING COUNTRY

For good reasons, cattle feeding has grown rapidly and reliably in this area.

Commercial cattle feeding in the Southwest has grown dramatically during the past 25 years. While the number of feedyards has decreased, feedyards have grown larger and the number of cattle fed has increased – to over 6 million head per year or about 25 percent of the U. S. total. That's why the TCFA area of Texas, Oklahoma and New Mexico is called "Cattle Feeding Country."

Here are the feedyards which in 1992 were TCFA Feedyard Members and a brief history of each, based on information supplied by each feedyard and information in TCFA files:

Abilene Cattle Feeders
Abilene, Texas

Built in 1964, capacity 6,000, last expanded in 1978 to 30,000. Original feedyard name, Davis Feedyard (1964-73); owner Edgar Davis; manager Alton Whittaker. Previous owners, Bob Morford, Jim Alexander, Robert Gooch and Bill Jay (1973-92). Current owner Alderman-Cave Feeds. Manager Bill Jay.

Agri Research Center, Inc.
Canyon, Texas

Built in 1980, capacity 240, expanded in 1986 to 3,000. Original feedyard name, Palo Duro Veterinary Services (1980-86). Owner David Bechtol, DVM. Manager Audie Waite.

Alfadale Stock Farm
El Reno, Oklahoma

Built in 1964, capacity 1,000, expanded in 1984 to 3,500. Owner W.E. (Bill) Jeffrey. Manager Peter W. Jeffrey.

Alta Verde Industries, Inc.
Eagle Pass, Texas

North yard built in 1966, capacity 2,500, expanded in 1973 to 25,000. Owner Leon Miller. Previous managers, Leon Miller, Newman Stutts, Glenn Polhemus, Ron Akin, Chip Miller, Bill Mies, Jonny Miller. South yard built in 1973, capacity 10,000, later expanded to 30,000. Original feedyard name, Maverick Beef Producers; originally owned by R.C. Harper. Later named Rio Grande Land and Cattle Co. (1977-81), owned by Rich Hormechea, Bob Rebholtz, Lee Bowen and Bob Hayes. Previous managers, Richmond Harper, Jr., Chad Foster, Rocky Reagan, Bob Hayes, Craig Clyburn, John Bliss.

B & F Feeders
Seminole, Texas

Built in 1973, capacity 2,000, expanded later to 6,000. Original feedyard name Cerrillos Cattle Co.; owners Pat Dunigan, George Ellis, Waldo Stephens (1973-76). Current owner and manager, since 1990, Gerald Franks.

Texas CattleWomen, organized in 1973, have steadfastly promoted beef – in supermarkets, in cookoffs and on the airways. (Left) Betty Carrothers, first president; (Right) CowBelles with KGNC Beefmobile on parking lot, ca. 1974.

Balmorhea Feeders, Inc.
Verhalen, Texas

Built in 1972, capacity 15,000. Previous feedyard owners, Joe Lee McMahon, Joe Lee McMahon, Jr. , David Lundrum, Charles Weinacht and Don Weinacht (1972-78). Previous managers, Charles Weinacht (1972-74), Nathan Reese (1974-76), Lonnie Winterrowd (1976-78). Current owners, since 1978, Charles Weinacht, Don Weinacht and Ronny Bowles. Manager, since 1978, Ronny Bowles.

Bar-G Feedyard
Summerfield, Texas

Built in 1966, capacity 12,000, last expanded in 1990 to 70,000. Original feedyard name, Alpha Omega Cattle Co., owned by Chuck Healey, Sam Huttenbauer, Jr., Leonard Meranus, David Gibson, Jimmy Brown and Milton Rudder. In 1968, became United Beef, owned by Sam Huttenbauer, Jr., Stanley Chesley and Milton Rudder; manager Milton Rudder. Current owners, since 1983, S.L. Garrison, Harvey Garrison, Stanley Schaeffer, Johnny Trotter, David Schaeffer, Eddie Sutton and Roy Bryan. General manager Johnny Trotter.

Barnes Feedlots, Inc.
Hobart, Oklahoma

Built in 1943, capacity 500, last expanded in 1972 to 3,500. Previous owners, Sid Barnes and J. D. Simms (1943-55); Sid Barnes (1955-72). Owner and manager (1972-92), George S. Barnes.

Barrett-Crofoot Feedlots – West
Hereford, Texas

Result of a merger of two feedyards: (1) Easley Feedyard, capacity 1,600, built in 1954 by Oscar Easley; and (2) Circle 3 Feedyard, capacity 30,000, built in 1967 by Joe Easley and John Pitman. Both feedyards sold in 1970 to John Pitman and were renamed Pitman West and Pitman East. Previous managers, Joe Easley (1959-67), Pitts Harrison (1973-76), Dean Stallings (1973-74), Milton Rudder (1977), Dennis Farley (1978), Logan B. McClelland (1979-80), Robert Barrett (1980-88). Current capacity 58,000; owners, Ed C. Barrett and Jay Crofoot. Manager Rodney Barrett.

Barrett-Crofoot Feedlots – East
Hereford, Texas

Built in 1969, capacity 28,000, last expanded to 42,000. Original feedyard name, Big Tex Cattle & Grain, owned by W. T. Carmichael, Kenneth D. Alban, Hilrey L. Aven and eight other investors (1969-77). Previous managers, James McCarty (1969-71), Bud Bowman (1972-75), Wayne Barber (1976), Ed C. Barrett (1977-91). Owners Ed C. Barrett and Jay Crofoot. Manager Robert Barrett.

Bartlett I
Tulia, Texas

Built in 1970, capacity 31,000, expanded in 1977 to 35,000. Previous feedyard name, Prairie Cattle Co., owned by 15 local stockholders. Previous managers, Frank Smith (1970-72), Sam Kirk (1972-77), Jerry Anderson (1977-79), Stan Price (1980-82), Jim Hazen (1982-83), Charley Cox (1983-84) and Mike Browning (1984-87). Current owner, Bartlett and Company. Bartlett Cattle Co.: Bob Hillier, President; Mike Gesling, manager.

Bartlett II
Hereford, Texas

Built in 1966, capacity 15,000, expanded in 1974 to 40,000. Original feedyard name, Hereford Cattle Feeders (1966-69), owned by Jim Bob Allison and Ray Polan (1966-69), manager Ray Polan. Previous feedyard name, Deaf Smith Feedyard (1969-90); owned by Bub Smith, Neil Yows and Bill Irwin (1969-70); owned by Southwest Investment Corp. (1970-78). Previous managers, Ernest Latta (1969-75), Bob Hillier (1975-78), Monte Garrett (1978), David May (1979), Stan Price (1980), Jim Ray (1981-83), Lyndon McLintock (1984-85) and Greg Weihrman (1986). Current owner, Bartlett and Company. Bartlett Cattle Co.: Bob Hillier, President; Mike Browning, manager.

Beefco Feeders, Inc.
Hereford, Texas

Built in 1942, capacity 400, last expanded in 1981 to 9,000. Original feedyard name, Beef City Feedyard (1942-64); owner, Jack Bradley (1942-89); manager, Jack Bradley (1942-76). Previous feedyard name, Cimarron Feedyard (1964-84). Previous owner, Ray Gerk (1989-90). Previous managers, Ray Gerk (1976), Greg Carlile (1981-86), George Keenan (1987-92). Current owner Jack Bradley. Manager Audie Baca.

Bennett Cattle Inc. Feedyard
Hart, Texas

Built in 1970, capacity 1,500. Previous name, Ed Bennett Feedyard, owned by Ed Bennett (1970-73). Current owners, Ed Bennett, George Bennett II, Harold B. Bennett, Mark Bennett. Manager George Bennett II.

CATTLEWOMEN: SUPER PROMOTERS

In 1973, when TCFA was six years old, the wives of cattle feeders decided to form a "partner organization." Betty Carrothers, wife of TCFA Charter President Jack Carrothers of Friona, took the lead and formed the Texas CowBelles.

At the organizational meeting in Amarillo, 21 women attended. They adopted the original mission of the American National CowBelles, formed in 1939, "to promote friendly and social relations among cattle people and to cooperate for the best interest of our industry, our community and our country."

From the start, ranch women joined the cattle feeding women, swelling the membership. By the end of 1973, the "yearling" had grown to 1,450 members with 12 local chapters.

Never short on ideas, the CowBelles took on such projects as Beef Gift Certificates, the Meat Identification Contest at the annual 4-H Roundup, booths at teachers conventions, distribution of nutrition materials to schools, Beef for Father's Day, in-store promotions, scholarship funds and – of special importance in recent years – legislative calling trees and the State Beef Cook-off.

In 1986, the name of the organization was changed to Texas CattleWomen, in line with the name change of the parent organization – to American National CattleWomen (ANCW).

The Oklahoma CattleWomen and the New Mexico CattleWomen have been equally active and effective. Texas has had one president of ANCW, Frances Owen (1985); Oklahoma has had three, Betty Smith (1960), Delores Finney (1967); Ann Woolley (1981); and New Mexico has had one, Dorothy Vaughan (1988).

Bogle Feedyard
Dexter, New Mexico

Built in 1937, capacity 1,000, expanded later to 12,000. Previous managers, Hal Bogle and Bill Bogle. Owner Bogle Farms. Manager Donald Bogle.

Boise City Feedyards, Inc.
Boise City, Oklahoma

Built in 1967, capacity 3,500, expanded in 1987 to 10,000. Original owners were 12 members of the Crabtree and James families (1969-79). Previous managers, Hank Ellison (1967-69), Terry Jennings (1969-70), Newt James (1970-71), Tarrol Cooper (1971-73), Don Bowers (1973-79). Current owners, Ron Overstreet, J. B. Stewart and Ross Biaggi. Manager, since1979, Ron Overstreet.

Bovina Feeders, Inc.
Farwell, Texas

Built in 1969, capacity 10,000, last expanded in 1987 to 42,500. Previous owners, D.C. Herring, Tom Caldwell, Troy Christian, Wendel Christian, Bob Wilson, Howard Fleming, Ralph Shelton, and Leon Grissom (1969-84). Previous managers, Wendel Christian and Leon Grissom (1969-71), Redge Priest (1971), Lee Hicks (1971-76), Gaylon Black (1976-79), David George (1979-81). Current owners, since 1984, Ralph Shelton and Leon Grissom. Manager, since 1981, Gail Morris.

Bradley 3 Ranch Feedlot
Memphis, Texas

Built in 1986, capacity 250, expanded in 1989 to 900. Owners, Bill and Minnie Lou Bradley. Manager Minnie Lou Bradley.

Bridges Feedyard
Hart, Texas

Built in 1970, capacity 1,500, expanded in 1987 to 4,500. Original owner and manager John Bridges (1970-90). Current owner and manager Bruce Bridges.

Brookover's Feedyard
Keyes, Oklahoma

Capacity 8,000. Owners Sam and Forest Brookover. Manager Robert Bergner.

Burner Cattle Company
Tornillo, Texas

Built in the early 1940s, capacity 6,000, expanded in 1960 to 12,000. Original feedyard name, Tornillo Feedyards; owner and manager, Bill Weathersbee (1940s-58). Previous owner and manager, Jess Burner, Sr. (1958-87). Current owner, Jess Burner, Jr. Manager John Orr.

C Bar Feedyard
Plainview, Texas

Built in 1955, capacity 3,000, expanded in the 1970s to 18,000. Original feedyard name, Carter-Kirchhoff Feedyard; owners, R. M. Carter and John Joe Kirchhoff (1955-1986); manager R.M. Carter (1955-83). Current owners, R. M. Carter family; leased, since 1986, to C Bar Feedyard, owned by C. C. (Cap) Naegle, Charles E. Koontz, Bill Carrothers and American Meat Trading Co. General manager R. M. Carter, manager Robert E. Carter.

CBC Feedyard
Friona, Texas

Built in 1964, capacity 1,500, expanded in 1973 to 4,500. Original feedyard name, London Feedyard, owned and managed by Richard London (1964-91). Currently leased to Dale Christie and Robbie Christie. Manager Robbie Christie.

Cal-Tex Feed Yard, Inc.
Trent, Texas

Built in 1966, capacity 40,000. Previous feedyard owners, R. L. Bland, Jack Gillis and Howard Foster (1966-68); Western Beef (1968-82). Previous manager, R. L. Bland. Current owners, the Bland family (Rosemary, Rex, Steve, John, Diltzie and Becky). Manager Rex Bland.

Canadian Feedyards, Inc.
Canadian, Texas

Built in 1972, capacity 17,000, last expanded in 1984 to 25,000. Original feedyard name, Canadian Feedyard–A Division of Beef Cattle Co. of Wheeler (1972-76). Previous owners, Dick Waterfield, Allen Webb, Joe Hathoot, Jack Watson, H. E. Wright (1976-83). Previous managers, Garland Rogers (1972-74) and Dick Waterfield (1976-83). Current owners, since 1983, Joe Hathoot, Allen Webb, the Webb children, Jack Watson and H. E. Wright. General manager Joe Hathoot, manager Kennith Thrasher.

Caprock Industries, Inc. #1
Gruver, Texas

Built in 1969, capacity 30,000. Previous owners, Ken Irwin, A. R. Borg, and a group of local investors (1969-70); Amarex Oil Co. (1970-71); Ken Irwin, A. R. Borg, James H. Jones and George Enloe (1972-74). Previous managers, George Enloe (1969-74), Doyle Cates (1975), Bill Park (1976-77), Johnny Posey (1977-81), J. D. Wilson (1981-86), Doug McCloy (1986-90). Current owner, since 1974, Cargill, Inc. Caprock Industries, Inc.: Zay Gilbreath, president; Bill Park, vice president of operations; Richard Winter, feedyard manager.

Caprock Industries, Inc. #2
Gruver, Texas

Built in 1970, capacity 28,000, later expanded to 32,000. Original feedyard owner, Ken Irwin, A. R. Borg and a group of local investors, later leased to Winrock Farms, then to Bill Foxley. Previous managers, George Enloe (1970-71), Marvin Wilson (1975-76), Farrell Johnson (1976-80), Bo Kizziar, (1980-82), Ray Etheredge (1983-88), Pat O'Connell (1988-92). Current owner, since 1974, Cargill, Inc. Caprock Industries, Inc.: Zay Gilbreath, president; Bill Park, vice president of operations; Jeff Rudolph, feedyard manager.

Caprock Industries, Inc. #4
Dalhart, Texas

Built in 1969, capacity 30,000, last expanded in 1986 to 54,000. Original name, Dalhart Feeders. Previous owners, Lee Bowen, Harold Daniels, Sam Singleton and Earl Wilbur (1969-71); Stratford of Texas (1971-73). Previous managers, Peyton Smith (1969-71), Gale Turner (1971-73), Marvin Wilson (1976-80), Farrel Johnson (1980-82). Current owner, since 1976, Cargill, Inc. Caprock Industries, Inc.: Zay Gilbreath, president; Bill Park, vice president of operations; Odis (Bo) Kizziar, general manager; Doug McCloy, operations manager.

Aledo Feedyard at Aledo, Texas was operated by Ray Smyth, 1950-77. Neiman-Marcus of Dallas included it on tours for foreign dignitaries.

Caprock Industries, Inc. #5
Bovina, Texas

Built in 1967, capacity 19,000, expanded in 1969 to 35,000. Original feedyard name, Parmer County Cattle Co. (1967-86); owners, Bunker Hunt, Bud Dillard and C. C. (Cap) Naegle (1967-84). Previous owners, O. L. Gragg, Jim Schwertner, Eugene Schwertner, John Byrum, Dick Waterfield and A. L. Black (1984-86); previous managers, C. C. (Cap) Naegle (1967-87), Pat Halverson (1987-89), Mike Heard (1989-92). Current owner, since 1986, Cargill, Inc. Caprock Industries, Inc.: Zay Gilbreath, president; Bill Park, vice president of operations, Steve Adams, feedyard manager.

Caprock Industries, Inc. #6
Lockney, Texas

Built in 1969, capacity 15,000, last expanded in 1987 to 35,000. Previous names, Lockney Feedyard (1969-78), Crofoot-Lockney, Ltd. (1979-80), Western-Lockney, Ltd. (1981-82). Previous owners, Lockney Co-op Gin (1969-84); Mexia Corp, leased to Western Beef, Inc. (1984-86). Previous managers, Ralph House (1969), Dolph Lusk (1970), Hank Winter (1971-86). Current owner Cargill, Inc. Caprock Industries, Inc.: Zay Gilbreath, president; Bill Park, vice president of operations; Mike Heard, feedyard manager.

Carrizo Feeders
Texline, Texas

Built in early 1970s, capacity 10,000. Original owner, Rockefeller Farms; Bob Beck, manager. Previous owners, Ron Offett, Circle H Farms. Current owners, Ray Snead, Bud Snead, Will Snead, Tom Hills, Douglas P. White. Manager Douglas P. White.

Carson County Feedyards, Inc.
Panhandle, Texas

Built in 1969, capacity 12,500, later expanded to 13,500. Owners, Frank and Mary Lou Simms, Marvin Sparks, Leslie McNeill and O.C. Holt. Previous manager, Frank Simms (1970-79). Current manager, since 1979, Leslie McNeill.

Cattle Management Systems, Inc.
Altus, Oklahoma

Built in 1969, capacity 10,000. Owned and managed by Paul Mock, John Mock and Don Kizziar.

Cattle Town, Inc.
Hereford, Texas

Built in 1969, capacity 21,500, last expanded in 1991 to 40,000. Previous managers, Hank Winter (1969-70), Walter Lewis (1970-71), Wade Lewis (1972) and Bill Teel (1973). Current owners, Fern Barnett, Chester Wiggains, Kenneth Christie, Marguerite Cole, Carrie Tatum, Wesley Barnett, Tommy Tatum, Bobby Owen, Rocky Lee and A.R. Dillard. General manager, since 1973, Lloyd Olson. Manager Alvin Downey.

Central Cattle Feeders
La Vernia, Texas

Built in 1972, capacity 4,000. Owner Fred Pierdolla. Previous manager Dick Moody (1984-87). Manager Jeffrey Pierdolla.

Central Plains Feedyard
Buffalo, Oklahoma

Built in 1963, capacity 2,500, last expanded in 1990 to 31,000. Original feedyard name, Buffalo Feedyard; owner, John Pitts (1963-76); manager, John Pitts (1963-65). Previous owners, Verlon Erwin (1977-88), Hank Walton and Bobby Rufenacht (1988-91). Previous managers, Ernest Latta (1965-70), Verlon Erwin (1970-87). Current owners, Hank Walton, Lloyd Hanahan and Jim Runyan. Manager Gene McGlaun.

Champion Feeders, Inc.
Hereford, Texas

Built in 1966, capacity 5,000, last expanded in 1972 to 32,000. Original feedyard name, Three R Feedyard, owned by E. C. Reinauer, Eddie Reinauer, Jr. and Bill Reinauer (1966-71). Previous managers, Dick Coupe (1966-67), Walter Lewis (1968-69) and Jim Flowers (1969-71). Previous owners, Southwest Feedyard and 20 local stockholders (1971-73); Adams Resources, Inc. (1973-77); Jess Burner, Sr., Charles Tow, Joe Mendiburu and Dave Hopper (1977-87). Current owners, Dave Hopper, Charles Tow, Joe Mendiburu and Gordon Dutterer. Manager, since 1971, Dave Hopper.

"Eat More Beef" signs dotted the Texas Panhandle during the 1950s. A few still stood 35 years later.

Welcome sign at Alta Verde Industries, Eagle Pass, 1970s and 80s. Alleys were named T-Bone Lane, Sirloin Street, etc.

Photos by Charles E. Ball

Chaparral Feeders, Inc.
Uvalde, Texas

Built in 1973, capacity 3,200, expanded in 1982 to 8,000. Original feedyard name, Chaparral Feedyard, Inc.; owner and manager Ray Manor (1973-75). Current owner and manager, since 1975, Ralph Hesse.

Chaparrosa Feedyard
LaPryor, Texas

Built in 1974, capacity 4,000, last expanded in 1986 to 7,500. Previous owner, B. K. Johnson Interests; B. K. Johnson and Dudley Campbell. Previous managers, Wayne Hamilton, Gus Sutton, Jim Jacobs, John Smithson and general manager Ed Stanfield. Current owner Chaparrosa Ranch. Manager Scott Thompson.

Cimarron Feeders
Texhoma, Oklahoma

Built in 1969, capacity 8,000, last expanded in 1987 to 46,000. Original feedyard owners, Dayle Oyler, Billy Ray Gowdy, L. J. Stafford, Roy Nall, Dave Burrow, Jerry Thrash, Vic Carey, Andy James, T. O. Puch and Julius Cox. Previous owners, Bill Gillenwaters, Mary Mag Nall, Gene Sizemore, J. B. Smith, Wayland Smith, Dub Jiles and Juanita Burrow. Previous managers, Bill Gillenwaters (1969-88), Ken Betschart (1988-90). Current owner, since 1987, Continental Grain Co. Manager John Rakestraw.

Circle Three Feed Yards, Inc.
Hereford, Texas

Built in 1970, capacity 3,500, last expanded in 1990 to 20,000. Originally, Polan Grain and Cattle Co., owned and managed by Ray Polan (1970-73). Later named Moorman Feedyard (1973-85), owned by Moorman Manufacturing, manager Frank Daniel (1973-85). Current owners, since 1985, Elmo Hall, Scott Hall and Jim Perrin. Manager Scott Hall.

Clayton Cattle Feeders, Inc.
Clayton, New Mexico

Built in 1969, capacity 10,000, expanded in 1973 to 16,500. Originally owned by a group of 25 stockholders (1969-81). Previous owners, W. Bascom Halcomb, Matt Irwin, Ethan Labrier, K. S. Eng, Clyde and Betty Sowers, William H. (Bill) Shaw, Charlie and Gloria James (1981-87); Charlie and Gloria James, William H. (Bill) and Verna Shaw (1987-90). Previous manager, Charlie James (1969-90). Current owner, William H. (Bill) Shaw. Manager Mark Polhemus.

Clint Feed Yard, Inc.
Clint, Texas

Built in 1954, capacity 500, last expanded in 1971 to 9,000. Original feedyard name, David Surratt Feedyard, owned by David Surratt. Current owners, since 1976, Fay Surratt Greenwood, Janice Armstrong, Jim Armstrong, Diane Otten and S. Brown. Manager, since 1956, Jim Armstrong.

Clovis Feed Yards, Inc.
Clovis, New Mexico

Built in 1955, capacity 5,000, expanded in 1984 to 11,000. Previous owners, G. W. Keenan, Lester Merrill, C. B. Erdwurm (1958-63); G. W. Keenan and C. B. Erdwurm (1963-70); C. B. Erdwurm (1970-83). Previous managers, G. W. Keenan (1958-70) and Rip Grinde (1970-77). Current owners, since 1983, C. B. Erdwurm Estate (Jim Lewter, Sharon Lewter, Brad Lewter, Tracy Clark and Gary Lewter). Manager, since 1977, Bob Tabb.

Cluck Cattle Company
Hart, Texas

Built in 1945, expanded in 1969 to 17,000. Previously owned by Jimmy and Roy Cluck (1945-70), managed by Jimmy Cluck (1945-82). Current owner Jimmy Cluck. Manager Mark Cluck.

Comstock Cattle Corp.
Amarillo, Texas

Built in 1961, capacity 2,000, expanded in 1970 to 2,600. Original feedyard name, B & C Cattle Co.; manager Sam Elliott (1961-87). Current owner Sam Elliott; Comstock Cattle Corp., lessee. Manager, since 1987, Dean Stallings.

Coronado Feeders
Dalhart, Texas

Built in 1973, capacity 35,000, last expanded in 1989 to 58,000. Original owners, a group of 15 local investors (1973-79). Previous managers, Jimmy Brown (1973-74), Ducky Gallo (1974-75), Leo Vermedahl (1975-79). Current owner, Continental Grain Co. Manager, since 1979, Richard Chumley.

Coyote Lake Feedyard, Inc.
Muleshoe, Texas

Built in 1973, capacity 5,000, expanded in 1989 to 15,000. Original feedyard name, Coyote Lake Feedyard, subsequently operated as Birdwell Cattle Feeders. Original owners J.E. Birdwell and family. Previous managers, Clay Birdwell (1974-83), Connie Robinson (1984-86), Eddie Derrick (1987-88). Current owners, E. W. Williams, Jr. and Harold P. (Bo) Brown, Jr. Manager Lee Britting.

D.G.C. Feeders, Inc.
Gruver, Texas

Built in 1969, capacity 8,500. Original feedyard owners, Bob Cluck, Donnie Thoreson, Mrs. D. G. Cluck, Dean Cluck (1969-84). Previous managers, Lee Carr (1969), Jim Clark (1969-70). Current owners, Monte Cluck, Randy C. Brown, Teresa Weldon and Dean Cluck. Manager, since 1970, Randy C. Brown.

Da-Pry Cattle Company
Stinnett, Texas

Built in 1990, capacity 1,000. Owner and manager Rob Sanders.

David Hill Cattle Company
Hereford, Texas

Built in 1951, capacity 1,000, later expanded to 3,500. Original feedyard name, Western Feedyards, owned by Homer M. Cogswell (1951-56). Previous owners, James McAndrews (1956-77), Walter Paetzold (1977-78), Elmo Hall (1978-90). Previous managers, Barry Cox and Scott Hall. Current owners, Chris and Charles Cabbiness; leased to David Hill. Manager David Hill.

Davidson Feed Pens
Pecos, Texas

Built in 1952, capacity 1,000, last expanded in 1978 to 12,000. Original owner, Paul C. Davidson. Previous managers, Walter Box (1967-72), John Benedict (1972-85). Current owner, Paul C. Davidson Estate. Manager Gerald Lewis.

Phyllis George received from TCFA a dozen steaks each month she was Miss America, 1971. Leon Miller of Eagle Pass (right) made the first presentation.

Photo by Ralph Leone

TCFA treated the 1986 Cotton Bowl teams with a steak dinner. Mike Bowles, TCFA president (left) with Doug Flutie, Heisman Trophy Winner from Boston College.

Dean Cluck Cattle Co., Ltd.
Gruver, Texas

Built in 1961, capacity 1,000, expanded later to 14,000. Original owner and manager, Dean Cluck (1961-85). Current owner, Dean Cluck Cattle Co., Ltd. Manager Monte Cluck.

Dimmitt Feedyard, Inc.
Dimmitt, Texas

Built in 1967, capacity 20,000, expanded in 1982 to 40,000. Previous owners, Bob McLean, I. B. Brandon, G. L. Willis, Jr., A. J. Kemp, Jr., Jim Elder and A. C. Hays (1967-82); Kenneth Rolston and George R. Bagley, (1982-88); George R. Bagley, Charlie Wales, Jay O'Brien, Dale Winders, George Sides, Monty Boozer, Glenn Deen, Ron Davenport and Wolfsky family (1988-90). Previous manager, Ike B. Brandon (1967-73). Current owners, Bill Scivally, George R. Bagley, Charles Wales, Monty Boozer, George Sides and Dale Winders. Manager, since 1973, George R. Bagley.

Double A Feeders, Inc.
Clayton, New Mexico

Built in 1972, capacity 16,000. Original feedyard name, New Mexico Cattle, Inc. Previous owners, Rex E. Reeves (1972-76); Pete Pascoe, Phil Stadtler, Rex Reeves, Taylor Lawrence and Jack Sparrowk (1976-79); Rex Reeves (1979-90). Previous managers, Mike Reeves (1972-75), Michelle Reeves (1975-76), Charles R. Carter (1976-79), Fred Poling (1979-84). Current owners, Gene Atchley and Forrest Atchley. Manager Gene Atchley.

Driskill Feedyard, Inc.
Sabinal, Texas

Built in 1969, capacity 3,000, expanded in 1982 to 6,000. Owner and manager George D. Driskill.

Dumas Cattle Feeders, Inc.
Dumas, Texas

Built in 1960, capacity 5,000, last expanded in 1978 to 22,000. Original feedyard name, Diedrichsen & Son (1960-68). Previous owners, Moore County Feeders, Inc. (1960-64); Diedrichsen & Son, Inc. (1964-68); Western Beef, Inc. (1968-79); Granada Corp. (1979-91), Manager, 1960-92, Howard G. Diedrichsen. Current owner Leo Jessen. Manager Jim Holland.

Everett Bros. Cattle Co.
Stratford, Texas

Built in 1955, capacity 500, expanded in 1984 to 2,500. Owned by Everett Bros. Manager, since 1955, Joe F. Everett.

Rex McAnelly (right) was president of the T-Bone Club during most of the 1970s. Shown here with T-Bone speakers Rep. Bob Price (left) and NCA President John Trotman, 1973.

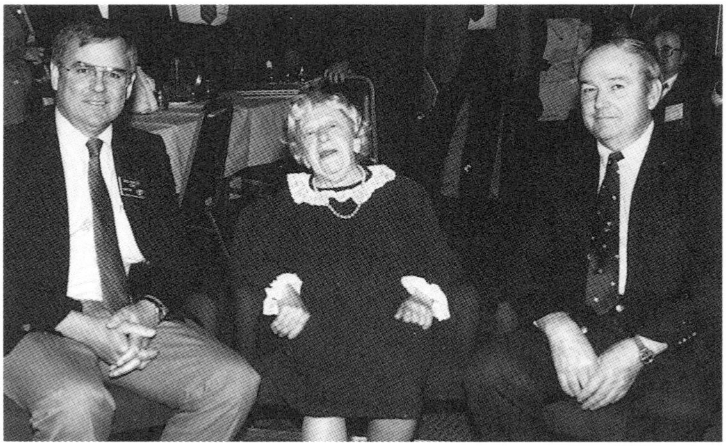

"Where's the beef?" Clara Peller, 82, made that a popular commercial for Wendy's hamburgers in the mid-1980s. Shown here with TCFA's Mike Bowles (left) and Bob Josserand, 1985.

Farris Feedyard
Hart, Texas

Built in 1989, capacity 1,250. Owner L. W. Farris, manager Ricky Farris.

Farwell Feed Yards
Farwell, Texas

Built in 1951, capacity 1,200, last expanded in 1984 to 30,000. Original feedyard name, Farwell Feedlot (1951-67); later changed to Far-Tex Feedyard (1968-74) and to Far-Tex Feeders, Inc. (1975-82). Previous owners, Jewell Castor (1951-59); Charlie and Bill Baxter (1959-64); Gilbert Schueler (1964-68); Agri Basic, Melvin Barton, James W. Witherspoon and Bill Waldrop (1968-74); Dean Stallings (1975-82). Previous managers, Pike Jordan (1951-60), Charlie and Bill Baxter (1960-64), Lee Hicks (1968-70), Dean Avara (1970-74), Johnny Hopkins (1975-80), Walt Olson (1980), Dean Stallings (1980-82), Phil Treadway (1982-91). Current owner AzTx Cattle Co. General manager Bob Josserand, manager Charlie James.

Flint Rock Land & Cattle Corp.
Gruver, Texas

Built in 1964, capacity 2,000, expanded in 1971 to 10,000. Original feedyard name, T. J. Harvey Feedyard (1964-74); owners, T. C. Harvey and Jack L. Gross (1964-74); manager T. C. Harvey (1964-74). Yard was closed 1974-85, then purchased by Flint Rock Land & Cattle Corp. Manager James W. Whitaker.

Freeman Feedlot, Inc.
Texhoma, Oklahoma

Built in 1951, capacity 5,000. Original feedyard owner and manager, M. H. Freeman (1951-59). Current owner and manager Jack M. Freeman.

Frio Feeders
Hereford, Texas

Built in 1981, capacity 3,000, expanded in 1985 to 6,000. Owners, Jim Nicholson and Jerry G. Curtis. Manager Jerry G. Curtis.

Friona Feedyard
Friona, Texas

Built in 1963, capacity 7,000, last expanded in 1967 to 28,500. Original owners, Jack Carrothers, G. B. (Pete) Buske, W. F. Buske, W. D. Buske, Paul Fortenberry, A. L. Black, Owen Seamands, Ranza Boggess, D. C. McWhorter and E. G. Tannahill (1963-74). Previous owners, Friona Industries (1974-85); Edwin L. Cox, Jr. (1985-90). Previous managers, Jack Carrothers (1963-73), Lee Britting (1973-86), Walt Olson (1986-88). Current owner, Friona Industries, L.P.; James Herring, president; Raymond Brown, general manager; David George, manager.

Gallo Feeders
Dalhart, Texas

Built in 1972, capacity 1,500. Original owner Jim Colquit. Current owners Al and Richard Gallo. Manager Richard Gallo.

Graham Land & Cattle Co.
Gonzales, Texas

Built in 1963, capacity 800, expanded in 1972 to 16,000. Original feedyard name, Harrell Cattle Co., owned by Wallace and Maurice Harrell (1963-87). Previous manager J. D. Gray (1963-86). Current owner Dr. Charles Graham. Manager Jay Gray.

Great Plains Cattle Feeders
Hereford, Texas

Built in 1972, capacity 2,000, expanded in 1985 to 16,000. Previous names, Texas "G" Feedyard (1972-79) and Texas Growers (1980-85). Previous owners, Iowa Agri-Nomics (1972-79); Sam and Ralph Grigsby (1979-84). Previous managers Jim Nicholson (1979-80) and Conrad Martin (1981-84). Current owners Clay Birdwell and John Ball. Manager Clay Birdwell.

Great Plains Cattle Feeders South
Dimmitt, Texas

Built in 1968, capacity 12,000. Original feedyard name, Taylor Feedyard (1968-83); owner Elizabeth Herring Estate, trustee Jay Taylor; manager Joe Smith. Previous feedyard name, Birdwell Cattle Feeders (1983-86), leased by J. E. Birdwell (1983-86); manager David May (1983-84). Currently leased by Clay Birdwell and John Ball. Manager Ed Derrick.

H & H Feed Lot, Inc.
Roscoe, Texas

Built in 1959, capacity 1,500, last expanded in 1968 to 15,000. Original owners, Philip V. Haynes, Ray Hendricks and 26 other stockholders. Previous managers, Philip V. Haynes (1959-74) and Tom White (1974-78). Current owners, Randall (Cuppy) Graham and Darel V. Nay. Manager, since 1978, Randall (Cuppy) Graham.

Harlingen Feedyard
Harlingen, Texas

Built in the early 1950s, expanded in 1987 to 8,000. Original name, Har-Tex Cattle Co. (1950s-62), owned by John Abbott and Milton Clapp, manager John Abbott. Previous owners, Jay Lewis Boggus (1962-68), Har-Tex Cattle, Inc. (1968-90). Previous managers, Doss Kornegay (1962-66), Sterling Ward (1967-86), Jimmy Youngblood (1986-90). Current owners, Kenneth Brown and Har-Tex Cattle, Inc.; leased since 1990 to Harlingen Feedyard. Manager Charlotte Rice.

Cattle Feeding: An Economic Giant

Who would have dreamed that, in three decades, Texas cattle feeding would grow into a $11.5 billion business?

That's how much economists figure Texas feedyards, which marketed 4.8 million head in 1990, contributed to the economy. Add to that the 1.2 million head marketed by Oklahoma and New Mexico feedyards, and the figure comes to $14.3 billion for the TCFA three-state area.

The economists arrive at that "economic impact" by use of a multiplier factor of 2.82, multiplied by gross sales of $4 billion in Texas. Gross sales in the three-state area were $5 billion.

The Texas State Comptrollers Office gave these impressive breakdowns for the industry, during 1990: Feedyards spent more than $1.1 billion to purchase Texas-grown feeder cattle, plus $355 million for feeds, and $105 million for trucking. Feedyards hired about 5,000 regular employees and several hundred contract workers.

The Comptrollers Office uses a 7.73 multiplier when estimating how many total jobs are created by Texas feedyards. Thus, nearly 40,000 jobs depend on the cattle feeding industry.

Breaking these figures down, the typical feedyard with 20,000 head capacity would feed about 40,000 head per year worth (in 1992) approximately $34 million in sales. On the expenditures side of the ledger, approximately $26 million would go for feeder cattle; $5.4 million for feed, coming from 18,000 acres of farmland; $1.25 million for interest, to banks and other financial institutions; and $410,000 in direct payroll, to 21 people. About 10 semi-trucks a day roll into the feedyard or 3,660 trucks a year – 2,300 with grain, 480 with feeder cattle and 880 for fed cattle.

TCFA officers escorted U. S. Rep. Charles Stenholm, from Texas, on a feedyard tour in 1985. From left: Mike Bowles, President-Elect; Stenholm; Bob Josserand, President; Zay Gilbreath, Vice President.

United States Sen. Phil Gramm, from Texas, visited TCFA several times. In 1985, he conducted a news conference at Randall County Feedyard, near Amarillo.

Hartley Feeders, Inc.
Hartley, Texas

Built in 1973, capacity 1,000, last expanded in 1992 to 22,000. Previous feedyard owners, Robert Thompson, Allen Thompson, Bruce Thompson (1973-76); Bruce Thompson and Robert Thompson (1976-89); Bruce Thompson (1989-91). Previous managers, Jeff Purvines (1973-75), Bill LaRoe (1976-84), Bruce Thompson (1984-88), Jeff Matsler (1988-89), Pat White (1989-90). Current owners, Bruce Thompson, Bill Kruse, David Noble, Joe Parker and Jeff Purvines. Manager Jeff Purvines.

Henrichson Feedyard, Inc.
Artesia Wells, Texas

Built in 1965, capacity 2,000, last expanded in 1991 to 10,000. Original owner, F. D. Henrichson (1965-87). Previous managers, F. D. Henrichson (1965-84), Trey Bell (1984-85), Philip Williams (1985-87). Current owner, Henrichson Feedyard Corporation. Manager Curtis (Corky) Weeks.

Henry C. Hitch Feedlot, Inc.
Guymon, Oklahoma

Built in 1952, capacity 3,000, last expanded in 1977 to 45,000. Previous feedyard managers, Cecil Reedy (1952-75), Dr. Don Williams (1975-82), Bill Skaggs (1982-89). Current owners, Hitch Enterprises, Inc. Manager Rod Schemm.

Hereford Feedyard, Inc.
Hereford, Texas

Built in 1961, capacity 5,000, last expanded in 1971 to 50,000. Previous owners Paul Engler, Jim Witherspoon, Ed Reinauer, Joe Reinauer, Fred Reinauer and other local investors (1961-70); ProChemco, Inc. (1970-80); AZL Resources (1980-83); Tofco, Inc. (1983); Bob Josserand, Milton Rudder, Marion Baugh, Ed Longcope, Scott Spangler (1983-87); Milton Rudder, Bob Josserand (1987-91). Previous managers Paul Engler, Dick Jagels, Richard Crider, Lloyd Olson, Walter Lewis. Current owner AzTx Cattle Co. General manager Bob Josserand; manager, since 1980, Wade Lewis.

Heritage Beef Cattle Co. Inc.
Wheeler, Texas

Built in 1969, capacity 12,000, last expanded in 1991 to 30,000. Original feedyard name, Beef Cattle Company; owned by founder Joe R. Weatherly and five stockholders: Charles Wright, Dick Waterfield, John Chambers, Bill Nix, Dale Nix. Previous owner, 1978-85, Joe R. Weatherly. Owners, since 1985, Joe R. Weatherly and Heinz Prechter. Manager Rodney Weatherly.

Hill Land & Cattle Company
Hart, Texas

Built in 1962, capacity 7,000, expanded in 1972 to 20,000. Original owner, Homer Hill. Previous managers, Bob Hill (1962-81), Louis Reeder (1981-83). Current owners, Bob Hill, Dennis Hill and Cindy Hill Studdard. Manager Dennis Hill.

Hitch Feeders I, Inc.
Hooker, Oklahoma

Built in 1967, capacity 16,000, expanded in 1974 to 60,000. Original feedyard name, Master Feeders I, Inc. Previous owners, H. C. Hitch, Jr., Ralph Grounds, Sonny Grounds, Leo Winters and Clark S. Willingham. Previous managers, Sonny Grounds (1967-69), Glen Cantrell (1969-76), Russell de Cordova (1976-80). Current owners, Hitch Enterprises, Inc. Manager, since 1980, Chip Newell.

Hockley County Feed Yard
Smyer, Texas

Built in 1977, capacity 10,000, expanded in 1980 to 15,000. Previous owners, Daryl Anderson (1977-88), George Wilson (1988-91). Previous managers, Bill Skaggs, (1977-82), Nile Morris (1982-84), Gary Cain (1984-88), Max Wilson (1988-91). Current owner, Koch Industries–Matador Cattle Co. Manager Bob Hornberger.

Hood County Feeders
Granbury, Texas

Built in 1970s, capacity 3,500. Previous owners, Floyd Bullard, Danny Harris, C.R. Puttett, Tom Holmes, James M. Meyer; later James M. Meyer, C.R. Puttett, Danny Harris, C.D. Moline, Joe M. Lyle, Larry Stewart; later James M. Meyer, Gary Shaw, David Smith. Previous managers, Jimmy Tidwell and Scott Thompson.

Horseshoe Feeders
Rule, Texas

Built in 1967, capacity 1,500, expanded in 1975 to 5,000. Original name, Brazos Valley Feeders (1967-90); owner and manager Bass Powell (1967-75). Previous owner and manager Bud Kitchens (1975-90). Current owner, Morrison Cattle Co. Manager Dusty Morrison.

Howell Cattle Company
Premont, Texas

Built in 1986, capacity 2,500, expanded in 1990 to 5,000. Owners, Jesse and Laura Howell. Manager Jesse Howell.

NATIONAL FOOTPRINTS

Almost from the beginning, TCFA had a national perspective – for beef promotion, legislation and national leadership.

TCFA leaders resisted the temptation to promote Texas beef. Instead, as early as 1971, they began funding cooperative projects with the Beef Industry Council of the National Live Stock and Meat Board and continued major contributions to the BIC ($250,000 or more most years) until the checkoff became effective (25 cents in 1981, $1 in 1986). During those years, TCFA forwarded a total of over $2 million out of dues to the BIC.

Likewise, TCFA leaders believed that no one could represent cattle feeders in Washington as well as our national organization. Consequently, major funding was directed to NCA – over $4.5 million in the last 25 years – making TCFA the largest financial supporter of NCA. This was the first association to pay NCA dues for all its members. In addition, TCFA officers and staff did an unusual amount of lobbying in Washington and during the 1980s hired top lobbyists (C. W. McMillan and Denise Bode) to assist NCA.

When the U.S. Meat Export Federation was formed in 1975, TCFA became a charter member and for several years maintained two memberships (now one) at $5,000 each.

The year 1982 was a summit year for TCFA on the leadership front. TCFA members served as the chief elected officer of three national organizations – Bob Carter as chairman of the Meat Board, George Feaster as chairman of the U. S. Meat Export Federation and W. J. Waldrip as president of NCA. (Although Waldrip was a Feeder Member of TCFA and later served on the Board of Directors, he was nominated for the NCA post by TSCRA.) Another TCFA member and former president, Robert D. Josserand of Hereford, became president of NCA in 1989.

Texas broke the "5,000,000 fed" mark in 1984. Amarillo Mayor Rick Klein (left) and Chamber of Commerce President Richard Ware (right) helped TCFA President Bob Josserand celebrate.

Steaks from the 5,000,000th animal fed in Texas in 1984 were presented to Texas Govenor Mark White (right) by TCFA President Bob Josserand.

Photos by Ralph Leone

J and R Feeders, Inc.
Rio Grande City, Texas

Built in 1991, capacity 1,000. Owners Joel F. Cortez, Jr. and Robert L. Villarreal. Manager Joel F. Cortez, Jr.

Jade Cattle Feeders, Inc.
Perryton, Texas

Built in 1973, capacity 4,000, expanded in 1985 to 7,000. Original feedyard name Dudley's Lot (1973-75), changed to Dudley's Feedlot Co. (1976-89); owner Eldon Dudley; manager Jim Hagg. Current owner and manager, Jim Hagg.

Keeling Cattle Feeders
Hereford, Texas

Built in 1963, capacity 5,000, expanded in 1990 to 14,500. Previous feedyard names, Sheepfeeders, Inc. (1963-68), Summerfield Feedyard (1969-75), Cheyenne Feedyard (1975-84). Previous owners, Marsh Pitman and Pitts Harrison (1969-75), Moorman Manufacturing Co., (1975-84). Previous managers, Tom Smith and Frank Daniel. Current owners, Scott Keeling and I. J. Tow. Manager Scott Keeling.

Kem Carr Preconditioning Pens
Canadian, Texas

Built in 1987, capacity 500, expanded in 1992 to 1,000. Owner and manager, Kem Carr.

Keystone Feedyard
Pearsall, Texas

Built in 1987, capacity 1,000, expanded in 1991 to 5,000. Owners, Philip and Rita Williams, Ray Alaniz and Ruby Alaniz Sotello. Manager Philip Williams.

King Brothers Ranch
Stanley, New Mexico

Built in 1951, capacity 1,200, last expanded in 1984 to 15,000. Owners, Sam King, Bruce King and Don King. Manager Bill King.

King Ranch Feedyard
Kingsville, Texas

Built in 1974, capacity 7,500, expanded in 1975 to 15,000. Original feedyard name, Eslabon Feedyard (1974-90). Owner King Ranch, Inc. Previous managers, Fred Radford (1974-78), Tom Harris (1978-89). General manager Tommy Haegelin, manager Richard Bretz.

King Ranch Cannonade Feedyard
Gonzales, Texas

Built in 1951, capacity 7,200. Previous feedyard names, J. S. Abercrombie Ranches, Inc. (1951-75) and Cannonade Ranch, Inc. (1975-91). Previous owners, J. S. Abercrombie (1951-75), J. E. Abercrombie (1975-91). Previous manager Tommy Haegelin (1977-88). Current owner King Ranch, Inc. General manager Tommy Haegelin; manager, since 1988, John Thiele.

Kingfisher County Feed Yards, Inc.
Okarche, Oklahoma

Built in 1969, capacity 10,000, expanded in 1973 to 15,000. Previous managers, Randy Hamilton (1982-84), Gary Van Horn (1984-87). Current owners, Marion Perdue and John McCarthy. Manager Chris Chitwood.

Kirkland Feedyard
Vega, Texas

Built in 1973, capacity 800, later expanded to 8,500. Original name, Scott Feedyard, owned and managed by James R. (Pedro) Scott (1973-82). Current owner and manager, Perry Kirkland.

LS Feeders, Inc.
Hart, Texas

Built in 1972, capacity 10,000, expanded in 1991 to 17,000. Original feedyard name, Hart Feeders (1972-77), later LS Feedyard (1977-89). Previous owners Harold, David and Danny Smith (1972-77); Bill Simpson, Marvin Simpson and Gene Lewis (1977-82); Bill Simpson, Gene Lewis and Louis Hall (1982-85); Louis Hall, Frank Sitton and Ron Sitton (1985-89). Previous managers, Jim Witt (1977-80), Buck Wehrbein (1980-82). Current owners, Bob Travis, Jim Lowder, Bob Hatcher, Mitch Johnson and Bert Gentry. Manager, since 1982, Bob Travis.

La India Feedyard
McAllen, Texas

Built in 1986, capacity 300, last expanded in 1990 to 4,400. Owned by A. A. Munoz II. Previous manager Joel Cortez (1986-89); current manager Eddie Guerra.

Lazbuddie Feeders
Lazbuddie, Texas

Built in 1975, capacity 700, last expanded in 1985 to 12,000. Original feedyard name and owner, J. P. Family, Inc. (1976-88). Current owners, Ralph Shelton and Leon Grissom. Manager, since 1976, Randy Thomas.

Live Oak Feedlot, Inc.
Batesville, Texas

Built in 1972, capacity 3,500, expanded in 1990 to 6,000. Owner and manager, W. J. Vaughan.

Livestock Industries, Inc.
Frederick, Oklahoma

Built in 1960, capacity 3,500, expanded to15,000, reduced later to 2,000. Original feedyard name, Coke and Quam Feedyard, owned by L. D. Coke and John Quam (1960-79). Previous managers, Junior Kelly (1960-79), Corky Walden (1979-87). Current owners, Kenneth and Peggy O'Hanlon. Manager Kenneth O'Hanlon, DVM.

Lone Oak Cattle Company, Inc.
Stephenville, Texas

Built in 1988, capacity 2,000, expanded in 1992 to 2,500. Owner and manager Duane Jones.

Lone Star Feedyard, Inc.
Happy, Texas

Result of a merger in 1974 of two feedyards: (1) Happy Wheat Growers, built in 1969, capacity 25,000; and (2) 7 Bar Cattle Co., built in 1972, capacity 20,000; resulting in a combined capacity of 45,000. Previous owners, Happy Wheat Growers Cooperative (1969-74); A. J. Wood, Dave Rhone and A. W. Lair (1974-75); Leon Miller and F. Lee Hicks (1975-76). Previous managers, A. J. McCarley, Wayne Bedwell, F. Lee Hicks (1975-80), Jeff Purvines (1980-91) Current owner F. Lee Hicks, manager Bill Kimmell.

Lubbock Feed Lots, LLP
Lubbock, Texas

Built in 1955, capacity 20,000, expanded from 1967-71 to 50,000. Original name, Lewter Feed Yard; owners, Murchison Brothers and D. W. Lewter; manager D. W. Lewter (1955-66). Previous owners, Earl Brookover and E. C. Crofoot (1966-69). Previous managers, Paul Morgan (1966-67), Ed Barrett (1967-68), Bass Powell (1968-69), E. C. Crofoot (1969-71), Jay Crofoot (1971-86). Current owners, Terry Crofoot, Jay Crofoot and Crofoot Family Trusts. Manager Terry Crofoot.

Ross Perot (right) received the TCFA Brave Bull, from Charlie Ball, following his speech at the 1975 TCFA Convention.

Govenor John Connally (left) received the TCFA Brave Bull, from Bob Bliss, following his speech at the 1980 TCFA Convention.

Luckey Custom Feedlot, Inc.
Devine, Texas

Built in 1972, expanded in 1988 to capacity of 15,000. Original feedyard name, Cox Feedlot (1972-85); owner Kenneth Cox; manager Philip Williams. Current owners, Bob Hayes, Kenneth Miller, Roger Rodante, Greg Hayes and J. H. Uptmore and Associates. Manager Bob Hayes.

Lykes Bros. Inc. Feed Yard
Edroy, Texas

Built in 1960, capacity 5,000, last expanded in 1987 to 22,000. Owner Lykes Bros, Inc. Previous manager, Hart Kornegay (1960-90). Current manager Robert J. (Rob) Kornegay.

Mc6 Cattle Feeders, Inc.
Hereford, Texas

Built in 1988, capacity 2,000, expanded in 1992 to 12,000. Owner and manager, Dan McWhorter.

McCasland Farms, Inc. Feedlot
Goldthwaite, Texas

Built in 1972, capacity 350, expanded in 1984 to 990. Owners, Homer McCasland and DeWayne McCasland. Manager Danny Sutherland.

McCloy Feedlot
Morse, Texas

Built in 1953, capacity 500, last expanded in 1984 to 12,000. Original feedyard owners, Mrs. M. W. McCloy and Sons (1953-77). Previous managers, Carson McCloy (1953-60), Wilson McCloy (1960-77). Current owners, Wilson McCloy, Bill McCloy, Mark McCloy, Allen McCloy and Nanelle Kiser. Manager, since 1977, Allen McCloy.

McDonald Bar 6
Mathis, Texas

Built in 1984, capacity 900. Owner and manager Jim McDonald.

McLean Cattle Company, Inc.
McLean, Texas

Built in 1973, capacity 20,000. Previous owners, Jake Hess, George E. Eck, Jake Hess II, Edward A. Patman (1973-78); Schmid Properties, Inc., George E. Eck and Edward A. Patman (1978-88); Byron N.Gossett, George E. Eck, Bruce N. Gossett and Danny J. Cobb (1988-92). Previous manager George Eck (1973-92). Current owners, Canadian Feedyard, Inc., Jay O'Brien, Tom Bivins, Teel Bivins and Mark Bivins. Manager Neal Odom.

McLeod Farms, Inc.
Morse, Texas

Built in 1978, capacity 600, expanded in 1988 to 3,500. Current owners, Thane McCloy, Sue McCloy and Rex McCloy. Manager Rex McCloy.

Mason Feeders, Inc.
Mason, Texas

Built in 1971, capacity 8,000. Original owner and manager, Marvin McMillan (1971-78). Current owner and manager, Lee McMillan.

Mayo Cattle Co.
Byers, Texas

Built in 1962, capacity 6,000, last expanded in 1975 to 13,000. Previous names, Clayco Cattle, Inc. (1962-88) and Red River Feedyard (1988-91). Previous owners, Joe J. Parker and J. H. Mayo, Sr. (1962-88); a group of 20 investors (1988-91). Previous managers Wendell Mayo (1962-88), Rod Crow (1988-91). Current owners Ray, Guy and Wendell Mayo. Manager Wendell Mayo.

Mesa Livestock, Inc.
Clovis, New Mexico

Built in 1950, capacity 4,000, expanded in 1986 to 7,000. Previous feedyard names, Bogle & Hance (1950-60), Bogle & Bogle (1960-79). Previous owners, J. D. Hance, Hal Bogle and Tom Bogle. Previous managers, Larry Tindell (1979-86) and Ronnie Tindell (1987-91). Current owners, Harvey Tindell, Larry Tindell and Ronnie Tindell. Manager Larry Tindell.

Moody Farms Feedlot
Pampa, Texas

Built in 1960, capacity 2,000, expanded in 1967 to 12,000. Owner Eugene B. Smith, Jr. Manager Rex McAnelly.

Morales Feed Lots, Inc.
Devine, Texas

Built in 1958, capacity 1,600, last expanded in 1990 to 12,000. Owner Pete Morales, Sr. (1958-75); manager Pete Morales, Jr. Current owners Pete Morales, Jr. and Ernie Morales. Manager Pat Seifert.

Morris Cattle Company
Pearsall, Texas

Built in 1957, capacity 500, expanded in 1990 to 900. Previous owner, Pat Morris (1957-83). Current owners, Pat Morris and Tobin Morris. Manager Pat Morris.

Morris Stock Farm
Gruver, Texas

Built in 1962, capacity 200, last expanded in 1991 to 6,500. Owner and manager Vance B. Morris.

N.A.P.I. Feedyard
Farmington, New Mexico

Built in 1982, capacity 10,000, expanded in 1990 to 14,000. Owner Navajo Agricultural Products Industry. Previous manager Jerry Anderson (1982-83). Current manager LoRenzo Bates.

Nortex Feedlot Company, Inc.
Dalhart, Texas

Built in 1956, capacity 5,000, expanded in 1972 to 17,500. Original name Boone Feedlot; owner and manager Paul Boone (1956-66). Previous managers, James McCarty (1966-68), Ducky Gallo (1968-70), Tommy Buckner (1970-80). Current owners, Arty Watkins, Charles Steel, Johnny Summerour and Bob Podzemny. Manager Shuck Donnell.

Oppliger Feedyard, Inc. South
Farwell, Texas

Built in 1964, capacity 2,500, expanded in 1988 to 17,000. Original name, Castor Feedyard (1964-82); owner Jewell Castor; manager Pike Jordan. Current owner and manager Don Oppliger.

Oppliger Feedyard, Inc. North
Clovis, New Mexico

Built in 1974, capacity 26,000. Original feedyard name, American Cattle Feeders (1974-91). Previous owners, E. F. Hutton (1974-76), Granada Corporation (1977-91). Previous managers, Robert E. Beck (1974-76) and Lee H. Stampe (1976-91). Current owner and manager, Don Oppliger.

Owen Bros. Custom Feeding Co.
Morton, Texas

Built in 1969, capacity 25,000, expanded in 1972 to 36,000. Original owners, Wayne Owen and Joe Hathoot (1969-70); Wayne Owen (1970-73). Previous managers, Tom White (1969-71), Don Leonard (1971-72), Ed Bricker (1972-75), Wayne Owen (1975-83). Current owners, Robert Musselman and Delton Caddell. Manager Robert Musselman.

PACO Feed Yard
Friona, Texas

Built in 1969, capacity 20,000, expanded in 1984 to 35,000. Original owners, Jack Carrothers, G. B. (Pete) Buske, W. F. Buske, W. D. Buske, Paul Fortenberry, A. L. Black, Owen Seamands, Ranza Boggess, D. C. McWhorter and E. G. Tannahill (1969-74); Friona Industries, Inc. (1974-85); Edwin L. Cox, Jr. (1985-90). Previous managers, John Gay (1969-70), Bill Teel (1970-73). Current owners, Floyd Reeve, Gary Brown, Jayson Grimsley, David Grimsley, Ron Davenport, Wesley Barnett, Tommy Tatum and Dennis Anthony. Manager, since 1973, Wayne Tunnell.

Palo Duro Feeders, Inc.
Gruver, Texas

Built in 1965, capacity 4,000, last expanded in 1987 to 32,000. Original name, Rafferty Feedlot (1965-69), later Palo Duro Feeders, Inc. Previous owners, Ed Rafferty (1965-69); Dale Hoover, O. C. Holt, Frank Pittman and Bill Pittman (1969-82). Previous managers, Ed Rafferty (1965-77), Penny Brooks (1977-80), Jim Dooley (1980-84), Walt Olson (1984-86), Craig Clyburn (1986- 88). Current owners, Texas Beef Producers Group; (W. H. O'Brien, Blake O'Brien, Alex O'Brien, Katherine O'Brien, John M. Shelton III, Malcolm Shelton, Jim Shelton, W. H. Attebury, Billy Attebury and Ed Attebury). General manager Craig Clyburn; manager Dwayne Thompson.

Pan Blanco Feed Yard
Raymondville, Texas

Built in 1967, capacity 150, later expanded to 450. Owner and manager Jim Mungerson.

Parker Cattle Company
Happy, Texas

Built in 1958, capacity 1,000, expanded in 1964 to 5,000. Original owner and manager J. O. Parker (1958-72). Current owner and manager Foster Parker.

Perry Feeders, Inc.
Pleasanton, Texas

Built in 1954 by D. T. Perry; operated until 1981 as a dairy and feedyard; capacity 3,000. Owners, Edna Ertel, Charlene Staha and C. J. Ertel. Manager C. J. Ertel.

Perryton Feeders, Inc.
Perryton, Texas

Built in 1969, capacity 26,000. Previous owners, Grady Yeary, Robert Urban, Dale McLain, Vern Conner, Lawrence Ellzey, H. C. Brown, Jr., Frank Buzzard, Billy Conner, Kermit Johnson, Roy L. Johnson, Tom Slack, Dean Slaughter, Byron Tevis, Cecil Tevis and A. B. Holmes (1969-79); Virl LaMunyon, Edgar Blair, David Warren and Billy Riley (1980-88). Previous managers, Jim Jacobs (1969-70), Glen Feldman (1970-72), Roy Johnson (1972-76), Chick Burney (1976-79), Boyd Colbert (1979-83), Lonnie Winterrowd (1983-86). Current owner Virl LaMunyon. Manager Ernest Latta.

Pride Feeders, Inc.
Hooker, Oklahoma

Built in 1972, capacity 24,000, expanded in 1988 to 38,500. Owners, Ralph and R. G. Grounds. Previous managers, Loyd Bennett (1972-74) and Merle Johnson (1974-80). Manager, since 1981, Doug Duncan.

Puckett Farms, Inc.
Sayre, Oklahoma

Built in 1962, capacity 500, expanded in 1975 to 1,000. Owners Blaine Puckett and Calvin Puckett. Manager Blaine Puckett.

Rafter 3 Feedyard, Inc.
Dimmitt, Texas

Built in 1984, capacity 11,000, expanded in 1991 to 37,000. Owners, Mike Baca, Gene Bradley and Andy Rogers. Manager Mike Baca.

Rafter O Cattle, Inc.
Canadian, Texas

Built in 1988, capacity 2,500, expanded in 1991 to 4,500. Owners, Grant and Marjorie Long. Manager Grant Long.

Photo by Joe Munroe

In a modern custom feedyard, each animal is checked every day for health. When a problem animal is spotted, it is taken to the hospital.

At the hospital, sick animals are treated by lay doctors, under supervision of consulting veterinarian. David T. Bechtol, DVM (left).

Rafter O Ranch Feedyard
Vega, Texas

Built in 1966, capacity 1,500. Owners Jay and Imogene Taylor, manager Leroy McCracken. Current owner, TORY Land & Cattle L.P., owned by Steve Taylor, Jane O'Toole, Penny Rembe, Sally Young. Manager Wes Mason.

Ranchers Feed Yards, Inc.
Fort Stockton, Texas

Built in 1958, capacity 4,000, expanded in 1973 to 17,000. Previous owners, Bert Kincaid, Jr. and Elton Holland, Jr. (1958-70); Elton Holland, Jr. and John Berry (1971-80). Previous managers, John Berry (1967-80), Elton Holland, Jr. (1980-88), Bland Holland, (1988-90), Pat Pearce (1990-91). Current owner and manager Elton Holland, Jr.

Randall County Feedyard
Amarillo, Texas

Built in 1967, later expanded to 65,000. Original owners, Tom Herrick, Jim Man, Wallace Locke, Gene McCartt, Oliver Bivins, Bryce Beard and Bruce Lusk (1967-69). Previous owners, Mesa Agro (1969-74) and Edwin L. Cox, Jr. (1974-90). Previous managers, Tom Herrick (1967-69), Bruce Lusk (1969-74), Rodney Herring (1975), Bill Ploesser (1976-79), Dwight Gleghorn (1979-81), Gene Pool (1981-82), Jim Hickman (1982-84), Buck Wehrbein (1984-91). Current owner, Friona Industries, L.P.; James Herring, president; Raymond Brown, general manager; Kevin Thompson, manager.

Reeves County Feeders, Inc.
Verhalen, Texas

Built in 1970, capacity 6,000, expanded in 1973 to 15,000. Previous owners, Tommy Mayfield (1970-75), Bill Ramsey (1970-87), Kees Vesheul (1975-87). Current owners, Scott, Paul and Dusty Evans, Elaine Evans Shock and Lori Evans. Manager Scott Evans.

Reuter & Reuter, Inc.
El Reno, Oklahoma

Built in 1985, capacity 1,750. Owner and manager Fred Reuter.

Reynolds-Wilson Cattle Feeders
Carrizo Springs, Texas

Built in 1971, capacity 3,000, expanded in 1989 to 13,000. Previous owners, T. A. Reynolds, Jr. and Herman Wilson Lumber Co. (1971-73), Reynolds-Wilson Lumber Co. (1973-80), TR Management, Inc. (1980-85). Previous managers, Roy Williams (1971-72), Dale Kizer (1972-73) and Tommy Reynolds (1973-85). Current owner Carrizo Ranch Corp. Manager Bob Owens.

Rio Beef Feedyard
Edinburg, Texas

Built in 1989, capacity 1,250, expanded in 1991 to 2,850. Owner Risica Family. Manager Mike Risica.

Rio Hondo Cattle Co.
Rio Hondo, Texas

Capacity 1,000. Owner Kenneth Otts. Manager Arlie Otts.

Saddle Mountain Land & Cattle Co.
Clayton, New Mexico

Built in 1962, capacity 5,000. Original feedyard name, Jesko Farms; owner Joe McDade; manager Bill Watson (1974-89). Current owner, since 1974, Norman Blankinship. Managers Chester and Linda Byrd.

San Angelo Feed Yard, Inc.
San Angelo, Texas

Built in 1962, capacity 5,000, expanded in 1990 to 15,000. Original owners, John Cargile family, Wade Choate family and Jack Drake (1962-86). Previous managers, Ron Appleton (1962-64), James McCartney (1964-67), Nile Morris (1967-72), Roy Williams (1972-81), Rod Crow (1981-86). Current owners, John Cargile family, Wade Choate family and Glenn Polhemus. Manager Glenn Polhemus.

Schmitz Feedlot
Clayton, New Mexico

Built in 1973, capacity 2,500, expanded in 1987 to 3,000. Original feedyard name, Mock Feeders, owned and managed by Keith Mock (1973-80). Current owner and manager Roger Schmitz.

Schumacher Feedyard
Hereford, Texas

Built in 1973, capacity 1,200, expanded in 1980 to 2,000. Previous feedyard names, Frio West Feeders (1973-78), Bar-G Cattle Co. (1979-83). Previous owner Frank Robbins (1973-84). Previous managers, Frank Robbins and Johnny Trotter (1977-83). Current owners, John Schumacher, W. J. Schumacher and Avis Blakey. Manager John Schumacher.

Seven X Cattle Feeders, Inc.
Summerfield, Texas

Built in 1968, capacity 24,000. Original feedyard name, Pre-Feeders, Inc. Previous owners, Bud Thurber, Jim Waterfield, R. Michael Robinson, J. D. Thompson, P. U. Howard, F. H. McQiddy, Thomas A. Harris and E. Hobart Fatheree (1968-83); Max Woodard, Mike Bowles and Jack Gilles (1983-88). Previous managers, Bud Thurber (1968-78), Mike Bowles (1978-86), John Van Pelt (1986-87), Charley Wilson (1988-89), Bill Holland (1989-90). Current owners, Brent Caviness, Terry Caviness, Kenneth Christie, Charles Hoover, Tom Jessop, Bill Thorne, Wayne Jordon, Harold Clark and Roger Jordon. Manager Stan Isaacson.

Seven Rivers Cattle and Commodities, Inc.
Carlsbad, New Mexico

Built in 1963, capacity expanded by 1990 to 30,000. Original feedyard name Seven Rivers Cattle Company; owners Gil Moutray and partners (1963-88). Previous manager Gil Moutray (1963-75). Current owner and manager Mike Welch.

Shepard Feedlot
Hale Center, Texas

Built in 1955, capacity 1,800, expanded in 1972 to 12,000. Owner Grady Shepard. Previous manager, Grady Shepard (1955-65). Manager, since 1965, Pat Shepard.

Smith Cattle Feeders, Inc.
Hereford, Texas

Built in 1971 by W. O. Markley, capacity 2,500, expanded in 1977 to 8,000. Later named Beef City, owned by Ray Gerk and Don Oppliger; 3-K Feedyard, leased by George Keenan; Riata Feedyard, owned by Jim Ward and James Ward. Previous managers, W. O. Markley, Ray Gerk, George Keenan and Jim Ward. Current owners, since 1989, Barry Smith, Robert Barrett, Rodney Barrett. Manager Barry Smith.

High moisture grain in trenches made news in 1967. Master Feeders at Hooker, Okla., put up 1¼ million bushels.

Mechanical loader for silage in trench, like this one at Moore County Feeders, Dumas, was unique in 1961.

Photos by Charles E. Ball

Southwest Feed Yards, Ltd.
Hereford, Texas

Built in 1969, capacity 42,000. Original owners Ralph Owens, Bill Gentry, S. L. Garrison, Charles Hoover, and 24 other local investors (1969-73). Previous owners, Bud Adams, Ada Resources and 12 minor stockholders (1973-79); Randy Best, Steve Adams, Bill Gentry. Ed Wick (1979- 83); Briarwood Manning Corp. (1983-88). Previous managers, Charles Hoover (general manager 1969-87), Bob Word (1969-73), Ken McLain (1974-81), Arnold Smith (1983-86). Current owners, Ben Murphy and Wayne Chastain. Manager Wayne Chastain.

Springlake Feedyard
Earth, Texas

Built in 1952, capacity 800, expanded in 1984 to 10,000. Original owner Ewing Halsell Ranch. Previous owner Edwin L. Cox, Jr. (1976-87). Current owner Friona Industries, L. P.: James Herring, president; Raymond Brown, general manager; Harold Greenhouse, manager, since 1976.

Spur Headquarters Ranch
Spur, Texas

Built in 1953, capacity 200, expanded in 1984 to 800. Owner O. J. Barron, Jr. Manager O. J. Barron III.

Starr Feedyards, Inc.
Rio Grande City, Texas

Built in 1974, capacity 14,000, expanded in 1986 to 20,000. Owners, Ken Scoggins, Jack Scoggins, Sr., Jack Scoggins, Jr., Lou Waters and W. B. Waters. Manager Jack Scoggins, Sr.

Stoney Point Feedyard
Melissa, Texas

Built in 1968, capacity 10,000. Original feedyard name, Meat Producers, owned by D. H. Byrd. Previous managers, Robert Cox, Eldon Merritt (1968-71) and Albert Fuller (1971-88). Currently leased by Clark Willingham, Mark Quinn, DVM and John Quinn. Manager Mark Quinn, DVM.

Sudan Livestock & Feeding Co.
Sudan, Texas

Built in 1941, capacity 1,000 cattle (plus some hogs), last expanded in 1988 to 23,000. Original owner Anton Rieder and heirs (1941-87). Previous managers, Ben Davidson (1949-67), Jim Davidson (1967-79) and Gery Donnell (1979-88). Current owners, since 1987, CDH Feeders, Inc. (Dwayne Harral, Dudley Harral, Glen Harral and Hap Squires). General manager Gery Donnell, manager Ron Akin.

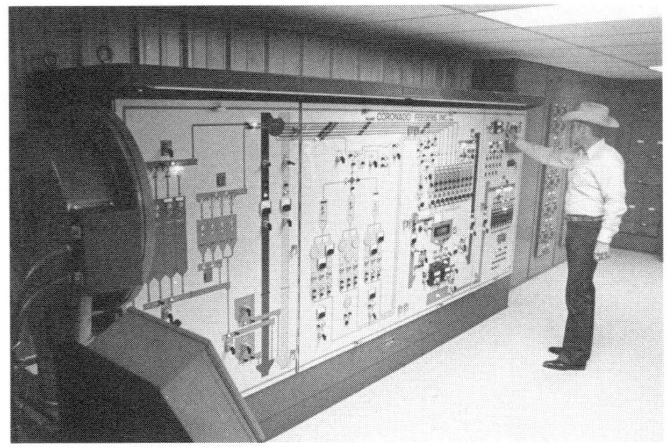

Heart of a high performing feedyard is a good feed mill . . . with electronic controls. Coronado Feedyard, 1975.

Sudan-East Feedyard
Littlefield, Texas

Built in 1974, capacity 25,000 expanded in 1978 to 30,000. Original feedyard name, Littlefield Feeders, owned by 63 local farmers and investors. Previous name, Littlefield Feedyards (1974-77), owned by Wayne Plowman (1974), Mondale Mills (1974-77), Edwin L. Cox, Jr. (1977-86); Friona Industries, L. P. (1986-92). Previous managers, Mike Thomas (1977-79) and Dave Koontz (1979-91). Current owners CDH Feeders, Inc. (Dwayne Harral, Dudley Harral, Glen Harral and Hap Squires). General manager Gery Donnell, manager Jeff Matsler.

Sugarland Feed Yards, Inc.
Hereford, Texas

Built in 1963, capacity 7,000; present capacity 27,000. Original name, Holly Sugar Feedyard (1963-68); owner Holly Sugar Corp. Previous owners, Melvin Cordray, Taft McGee, Henry Sears (1968-85). Current owners, Helen Burton, Ruby Kendrick Sears, Estate of James Sears, Bert Parker, Jan McElhaney, Roger Albracht and John Elliott. Former manager Melvin Cordray (1963-85). Manager Jan McElhaney.

Sunnyside Cattle Co.
Dimmitt, Texas

Built in 1961, capacity 1,000, expanded in 1991 to 12,000. Original feedyard name, Sunnyside Cattle Co.; owners, D. H. Schottenkirk, Ben Derrick and Buford Hayden (1961-68). Previous name, Diamond A Cattle Co. (1968-84); owner Robert O. Anderson; manager D. H. Schottenkirk. Current owner Mark Cluck. Manager Stanley Nelson.

Swisher County Cattle Co.
Tulia, Texas

Built in 1967, capacity 15,000, expanded in 1972 to 55,000. Previous feedyard owners, Hol Ed Toles, John Frank Toles, Vernon Harman, R. L. Bliss, B. R. Evans and V. H. Harman (1967-71); Mesa Agro (1971-74); Friona Industries (1974-85); Edwin L. Cox, Jr. (1985-90). Previous managers, R. L. Bliss (1967-74), Louis Reeder (1974-77), Mark Dunlap (1977-84), Ron Akin (1984-88), Kerry Teafatiller (1988-89) and Mike Bowles (1989-91). Current owner, Friona Industries, L. P.; James Herring, president; Raymond Brown, general manager; John Bliss, manager..

Tascosa Feedyard, Inc.
Bushland, Texas

Built in 1970, capacity expanded by 1976 to 15,000. Previous feedyard names, Producers Grain Corp.-West Plains (1970-80), West Plains Feedyard (1980-83), Playa Corp. (1983-85). Previous owners, Producers Grain Corp. (1970-80), J. Z. and Kathlene Smith (1980-83); Dick Coon, Richard Coon, Tom Doughtie and Sam Hill (1983-84); Dick Coon, Richard Coon, Tom Doughtie and Mark Dunlap (1985-90). Previous managers, J. Z. Smith, Moody Taylor (general manager), Paul Morgan (general manager), Andy Michaels, Jack Rhodes, Kem Rogers and Sam Hill. Current owners, Mark Dunlap, Randy Johnson, Virgil Boll, Rob Brown, Rob A. Brown, Jody Bellah. Manager Mark Dunlap.

Tejas Feeders, Inc.
Pampa, Texas

Owns and operates two feedyards. West yard built in 1966, capacity 20,000. Original feedyard name, Heaton Cattle Co.; owner Bill Stockstill (1966-72). Previous owners, ProChemco, Inc. (1972-80); Bill Stockstill (1980-86). Previous managers, Billy Joe McFall (1972-75), Bill Stockstill (1976-86). East yard built in 1972, capacity 15,000. Original feedyard name, Cockrell Cattle Company; owner Buddy Cockrell (1972-90). Leased to International Cattle Systems (1977-80), general manager Dale Lasater, manager Ray Etheredge; Canadian Feedyard (1981-83), manager Kennith Thrasher; S & C Feeders (1985-88), manager Glen Cantrell. Current owners Mike Smith and Roger Newman. Manager Leo Vermedahl.

Texas Beef Producers, Inc.
Dumas, Texas

Built in 1967, capacity 12,000, expanded in 1985 to 45,000. Original owners, W. H. O'Brien, Glenn Deen, Tom Marsh (1967-73); AZL Resources (1973-79). Previous managers, Glenn Deen (1967-83), Leo Vermedahl (1983-84), Bill McCoy (1985-88), Stewart Pike (1988-90). Current owners, Texas Beef Producers Group; (W. H. O'Brien, Blake O'Brien, Alex O'Brien, Katherine O'Brien, John M. Shelton III, Malcolm Shelton, Jim Shelton, W. H. Attebury, Billy Attebury and Ed Attebury). General manager Craig Clyburn, manager David Hutcherson.

Texas County Feedyard, Inc.
Guymon, Oklahoma

Built in 1964, capacity 9,500, last expanded in 1986 to 27,500. Original owners, Henry C. Hitch, Ralph Grounds, E. R. Kimsey, H. C. (Ladd) Hitch, Tom Shirley, Art Ebener, Sam Davis, R. G. (Sonny) Grounds and Leo Winters. Previous manager, E. R. (Ray) Kimsey (1964-76). Current owner, Brookover Companies. General manager E. R. (Ray) Kimsey, manager Dee Follis.

Texhoma Cattle Feeders, Inc.
Texhoma, Oklahoma

Built in 1957, capacity 2,000, expanded in 1990 to 4,500. Original name, Jones Feedyard. Previous owners, Elmo Jones (1957-77); Tommy and Larry Jones (1977-87). Previous managers, Elmo Jones (1957-77); Tommy Jones (1977-87). Current owner and manager Garland Dahl.

Texzona Cattle Feeders
Hale Center, Texas

Built in 1965, capacity 20,000, expanded in 1990 to 30,000. Previous feedyard names, PGC-South Plains Division (1965-78), South Plains Feed Yard (1979-84), Hale Center Feeders (1984-89). Previous owners, Producers Grain Corp. (1965-78); South Plains Feed Yard, Inc., owned by Pat Shepard, Howard Simpson, Waddell Strain, C. B. Hodge, Dick Lemond, Geral Miller, Jimmy Haggerton, Maurice Harrell, Kirk Morris and Hale Center Investments, Ltd. (1979-84); Hale Center Feeders, owned by Moody Taylor, Jan Tyler and limited partners (1984-89). Previous managers, Moody Taylor (general manager 1970-74), Paul Morgan (general manager 1975-79), Grover Phillips, Charlie Campbell, Charlie Christopher, Pat Shepard, Jim Witt, David May and Joe Bell. Current owners, Marvin Roberts, Mark Adams, Bob Rufenacht. Manager Pat Halverson.

Tri State Feeders, Inc.
Turpin, Oklahoma

Built in 1968, capacity 15,000, expanded in 1990 to 18,000. Original feedyard name, Panhandle Feedyard (1968-87). Previous owners, Lee Bowen (1968-86); Max Mendenhall, Robert Birney, Jack Cullison, J. I. Cullison and Raymond Powell (1986-89). Previous managers, Lee Bowen (1968-78), David Freelove (1978-88), Edward Morrow (1988-89), Bob Kellogg (1989). Current owners, Jack Cullison, J. I. Cullison and Ruby Powell. Manager Walt Olson.

Brahman crosses were discounted for years – until the 1970s when feedyards learned that up to one-fourth Brahman performed well and produced acceptable carcasses.

Photos by Charles E. Ball

Swisher County Cattle Co. has been known through the years as "a showplace feedyard." Built in 1967, capacity 55,000.

Tri-State Cattle Feeders
Hereford, Texas

Built in 1969, capacity 12,000, expanded in 1973 to 15,000. Original feedyard name, Tex-Sun. Previous owners, Carroll Newsom, R. N. Yarbro, Frank Barrett, Z. Cansler, Frank Zinser, Emory Brownlow, Joe Kerr, Dr. Hugh McCrary, Charles Owens, Rudy Metz and Cliff Skiles (1969-74); Tri-State Chemical and Bob Sims (1974-79); Bob Sims, R. N. Yarbro, Leo Helman, Sam Kirk, (1980-82); Bob Sims, Sam Kirk, Leo Helman (1982-89). Previous managers, Art Baker (1969-70), Wayne Barber (1970-74). Current owners, Bob Sims, Sam Kirk and Ron Banks. Manager Bob Sims.

Tulia Feedlot, Inc.
Tulia, Texas

Built in 1962, expanded in 1972 to 24,000. Original owners, Harvey Milner, Lewis Mahagan, R. K. (Pinky) Brooks, O. V. Morris, Mack Hale, H. T. Copeland, Roy Stockett, Arthur Jennings, Curtis Brasher, and Henry Boston. Previous managers Roy Stockett (1962-64) and Robert Allen (1964-89). Current owners, H. T. Copeland, Otis Harman, Charles B. Martin, Jr., N. M. Stockett, Morgan Sturgess, Mark Martin, Doug Martin, Roy Stockett, R & D Cattle, Indian Creek and Kathy Osburn. Manager John Van Pelt.

Union County Feedlot, Inc.
Clayton, New Mexico

Built in 1966, capacity 15,000, expanded in 1969 to 22,000. Previous owners, Rex E. Reeves, R. L. Bland, Jack Gillis, Howard Foster, Bill Armstrong, Ed Heringa and Gary Weisbart (1966-69); Western Beef Inc. (1969-84). Previous managers, Ralph Beatty (1966-67), Charlie James (1967-69), Bill McKenzie (1969). Current owner Virl LaMunyon. Manager, since 1969, Jack Butt.

Vaquero Cattle Feeders, Inc.
Floresville, Texas

Built in 1975, capacity 1,000, expanded in 1989 to 11,000. Owner K. R. Brown. Manager Charley Wilson.

Veribest Cattle Feeders, Inc.
Veribest, Texas

Built in 1963, capacity 4,000, expanded in 1968 to 12,000. Original feedyard owner and manager, Gene Newman (1963-77). Previous manager, Wes Bonner (1977-88). Current owners, Wes Bonner, Lad Linthicum, Ruth Weatherford, Ted Weatherford, Russ Weatherford. Manager Andrew Lewis.

Walter Lasley & Sons, Inc.
Stratford, Texas

Built in 1953, capacity 150, last expanded in 1983 to 18,000. Owners, Walter and LeVerne Lasley. Manager Walter E. Lasley.

Ward Feedyard
China Spring, Texas

Built in 1971, capacity 12,000. Previous name, Markum Cattle Co., owned by Burrell Markum. Current owner, since 1987, Buddy Markum. Manager Ivan Ward.

Wes Tex Feed Yards, Inc.
Muleshoe, Texas

Built in 1960, capacity 12,000, expanded in 1968 to 25,000. Original name, King Feed Yard, owned by Harold D. King (1960-84). Previous managers, Calvin Cooley (1960-62), Roy Weeman (fat cattle salesman 1962-74), Bob Bliss (1962-67), Bob Clark and Max King (1967-84). Current owners, Louis Reeder, Pete Bonds, J. D. Vann, Weldon Vann, Gary Vann, Roger Vann, Bob Clark and J. B. Douglas. Manager Louis Reeder.

West Plains Feedyard, Inc.
Friona, Texas

Built in 1967, capacity 40,000, expanded in 1973 to 65,000. Original feedyard name, Hi Plains Feed Yard (1967-87), owned by Lawrence Martin, Clarence Martin, Wesley Barnett, Ralph Shelton, Sloan Osborn, Joe Osborn and Charles Osborn. Previous managers, Paul Morgan (1967-72), Earl Elam (1972-75), Moody Taylor (1976), Bill Bailey (1977), Tom White (1978), and Clarence Martin (1979-87). Current owners, Carl Menke, Edwin Axe, Elvin Wilson, Richard Schlabs, Steve Hysinger and Bill McCoy. Manager Bill McCoy.

Wheeler Bros. Feedyard
Watonga, Oklahoma

Built in 1969, capacity 1,000, last expanded in 1990 to 20,000. Owner, Wheeler Bros. Grain Co. Previous managers, Buel Hoar (1969-71), Steve Smola (1971-85), Rick Sanders (1986-90). Manager Dan LaTourell.

Wheeler Feed Yard, Inc.
Wheeler, Texas

Built in 1972, capacity 12,000. Previous feedyard names, Prime Feeders, owned by Dick Hefley, T. A. Greenhouse, Bobby Deal, Harold Lloyd Lee, Dick Pugh, R. B. Shields, Chuck Hogan (1972-77), managers, Dick Hefley (1972-74), Chris Chitwood (1975-77); Far Tex Feeders, owned by Dean Stallings and a group of investors (1978-79), manager Glen Rabel (1978-79). Current owner Don McCasland. Manager, since 1979, Billy Joe McFall.

A fund-raising advertisement by the Animal Industry Foundation, 1991.

While some consumer groups protested, in the 1980s, TCFA and others pushed Beef Quality Assurance programs.

Woodley Ranch Feedlot
Sabinal, Texas

Built in 1972, capacity 5,000. Original feedyard name, Yucca Feedlot. Owner and manager, T. M. Woodley, Jr.

XCL Feeders, Inc.
Hereford, Texas

Built in 1973, capacity 1,250, expanded in 1988 to 4,000. Previous feedyard name, Walden Cattle Co., owner and manager John Walden (1984-90). Current owner Steve Lewis, DVM. Manager Jerry McConnel.

XIT Feeders
Dalhart, Texas

Built in 1970, capacity 28,500, last expanded in 1990 to 55,000. Original owners Oscar Easley, Joe Easley and 12 other investors (1970-73). Previous owner, Southwestern Investment Corp. (1974-78). Previous managers, Joe Easley (1970-73) and Bill Thorne (1973-86). Current owner, Continental Grain Co. Manager Les Howard.

Zavala Feeders, Inc.
Crystal City, Texas

Built in 1968, capacity 2,200, expanded in 1976 to 3,200. Previous feedyard names, Byrd Farm, Inc. (1968-76), Zavala County Feedlot (1976-91). Previous owners, D. H. Byrd (1968-76), Don Wetzig and Doc Hagan (1976-89). Current owners, Cleve Runnells, Jim Peters, Cleve Runnells, Jr., Don Wetzig, Mike Kirk and Mike Heller. Manager, since 1976, Don Wetzig.

"Together We Are Something!"

TCFA DIRECTORS

1967-1992

Name	Years Served
Alexander, James M.	1984
Allen, Robert	1967-74
Anderson, Anne Ilse	1986-88
Atchley, Gene	1990-92
Attebury, W. H.	1982-83
	1992
Bagley, George	1975-76
Ball, John R.	1989-91
Barnes, George S.	1984
Barnett, Wesley	1991-92
Barrett, Robert	1991-92
Barron, O. J., Jr.	1967-81
Bassett, Ron	1988-90
Baugh, Marion	1979-81
Bechtol, David T.	1987-88
Benedict, John N.	1980-82
	1986-88
Bibb, Jack	1992
Birdwell, Clay	1984-86
Birdwell, J. E.	1976-78
Bivins, Julian L.	1981
Bivins, Teel	1988-90
Black, A. L.	1971-76
	1988-90
Bliss, R. L. (Bob)	1970-83
Bonner, Wes	1982-83
	1987-91
Bowles, Mike	1983-88
Brandon, I. B. (Ike)	1970-73
Bratcher, Tom	1967
Brimhall, Van E.	1985
Britting, Lee	1992
Brown, Randy C.	1986-88
Brown, Richard	1987-89
Butt, Jack	1987-89
Carrothers, Jack	1967-73
Carter, R. M.	1967-75
	1981-83
Chumley, Dick	1986-88
Cluck, Bob	1970-74
Cluck, Monte	1990-92
Collier, Neal	1979-81
Cook, Jim	1980
Cooper, J. B., Jr.	1977-78
Cordray, Melvin	1982-83
Cox, Edwin L., Jr.	1976-77
Cox, Kenneth L.	1979-81
Crofoot, Jay	1975-77
Crofoot, Terry	1982-84
Davenport, Ron	1977-79
	1984-86
	1989-91
Davidson, Jim	1967
	1971-77
Deen, Glenn	1967-79
Denson, Kemp	1989-91
Diedrichsen, Howard	1987-89
Donnell, Gery C.	1985-86
Donnell, Shuck	1992
Drummond, Frederick	1991-92
Dudley, Eldon	1985-87
	1990-92
Dunlap, Mark	1989-91
Easley, Joe	1967-71
	1974
Eller, David G.	1980-82
Ellis, Jr., George F.	1968-74
Engler, Paul F.	1978-80
	1987-89
Enloe, George	1976-78
Farley, Dennis R.	1987-88
Fatheree, E. Hobart	1981-83
Feaster, George	1973-75
	1982-84
Felton, Scott M.	1983-85
Florence, Doug	1975-76
Follis, Dee L.	1989-90
Franks, Gerald	1983-85
Freeman, Jack	1969-74
Garrison, S. L.	1984-86
Geries, Dick	1987-89
Gilbreath, Zay	1979-81
	1985-89
Gillenwaters, Bill	1979-81
	1985-87
Gossett, Byron N.	1984-86
Gossett, Martin	1969-75
Graham, Randall	1986-88
Griffin, Mike	1981-82
Grissom, Leon	1986-88
Harman, Otis	1977-79
Harral, Dwayne	1991-92
Harrell, Wallace	1974-75
Harrison, Pitts	1976
Hathoot, Joe	1986-88
	1990-92
Hayes, Bob	1984-86
Haynes, Philip V.	1968-75
Herrick, Tom	1967-77
Herring, James E.	1991-92
Hesse, Ralph	1985-87
Hicks, Lee	1970-75
Hill, Dennis	1987-89
Hillier, Robert J.	1992
Hinkle, Rolla	1978-80
Hitch, H. C., Jr.	1978-80
Hitch, Paul	1975-77
	1991-92
Hoover, Charles	1975-77
Hopper, Dave	1978-80
	1989-91
Howard, Les	1989-92
Hunt, Jack	1981-83
Irwin, William K.	1983-85
Jagels, Richard	1967-73
James, Charlie	1983-85
Jarrell, John	1967
	1971-73
Josserand, Robert D.	1978-80
	1983-87
Kassahn, Jim S.	1985-87
Keeling, Scott	1989-91
Kemp, A. J., Jr.	1978-80
Kimsey, Ray	1976-78
King, Bill	1980-82
King, H. D.	1967-69
King, Max	1982-84
Kirk, Sam	1992
Koons, Lyle F.	1990-92
Koontz, Charles E.	1972-74
	1978-85
	1990-92
Kornegay, Hart	1976-78
LaMunyon, Virl	1983-85
Lasater, Dale	1980-81
Lasley, Walter	1967-76
	1983
Lasley, Walter E.	1987-89
Latta, Ernest	1974-75
Lemon, Glen E.	1981-82
Longcope, Ed	1980-82
Luckey, E. George	1967-68
Madden, Wales, Jr.	1979-81
Marsh, Virgil	1971-76
Martin, Clarence	1977-79
Martin, Quentin	1982-84
McAnelly, Rex	1974-75
McCasland, Don	1986-88
McCasland, Z A.	1968-69
	1971-76
McCleskey, Joe	1974
McCloy, Allen	1978-80
	1989-91
McCoy, Bill	1989-91
McDow, Ross	1991-92
McElhaney, Jan B.	1991-92
McMahon, Joe	1975-77
McMillan, Marvin	1975-77
McMurtry, J. Alfred	1968-70
McNeill, Leslie L.	1985-87
	1992
Middlebrook, S. B.	1967-68
Miller, Leon	1970-82
Mock, Paul	1969-75
Morales, Ernie	1990-92
Morford, Bob	1977-79
Moss, Rod	1981
Musselman, Robert	1987-89
Naegle, C. C. (Cap)	1967-75
	1981-82
Newman, Gene	1969-84
O'Brien, John Jay	1977-79
	1985-87
	1989-92
Ogle, Larry	1992
Olson, Walt	1992
Owen, Wayne	1968-76
Park, Bill	1982-84
Parker, J. O.	1967-71
Parker, Joe J.	1980-82
Parker, Robert L.	1981-83
Pitts, John, Jr.	1967-69
Powell, M. B.	1967-71
Priest, Redge	1980-82
Quinn, Mark	1990-92
Ragland, Bill	1982-84
Reeder, Louis C.	1984-86
	1990-92
Reuter, Gaylord	1981-83
Reuter, Jeffrey	1988-90
Reynolds, T. A., III	1983-85
Richter, John	1976-78
Rudder, Milton C.	1988-90
Samuel, R. E., Jr.	1976-78
	1992
Scarmardo, Pete	1991-92
Schwertner, Jim	1982-84
	1989-91
Scoggins, Jack	1980-82
Scoggins, Jack, Jr.	1988-90
Sears, Jim	1981
Shelton, John M., III	1976
Shepard, Pat	1971-76
Simmons, Tom B., Jr.	1967
Simms, Frank	1982-84
Simpson, F. Howard	1976-78
Sims, Bob	1985-87
Snead, Ray A., Jr.	1985-87
Stampe, Lee H.	1984-86
Summerour, Jimmy	1968-69
Summerour, Johnny	1967
Swift, Leon	1978-80
Tatum, Ellis	1979-81
Thorne, Bill	1977-79
	1983-85
Thurber, W. R.	1979-81
Trotter, Johnny	1988-90
Vaughan, W. J.	1991-92
Vermedahl, Leo D.	1978-80
	1983-85
	1988-92
Waldrip, W. J. (Dub)	1986-88
Waterfield, Dick	1974
	1978-80
	1982-86
Waterfield, Jim B.	1988-90
Watson, Farrell	1975-77
Weatherly, Joe	1977
Whittenburg, J. A., III	1968-74
Williams, Don	1982-84
Willingham, Clark S.	1985-90
Winter, Hank	1975-77
Winter, Richard A.	1984-86
Wohler, Jr., Wilson H.	1969-72
Woodward, J. Troy	1977-79
Worley, Victor	1977-79

APPENDIX B

TCFA EMPLOYEES

For One Year or More, By Date of Employment

Name	Position	Years Employed
Lloyd Bergsma	Executive Vice President	1967-72
Phoncille Driggers	Secretary	1967-68
Cathryn Covey	Market Reporter	1967-70
Joe Van Zandt	Assistant EVP	1969-71
Mildred Reeves	Office Manager	1969-74
Sherrian Rea	Secretary/Receptionist	1970-72
Bud Patterson	Information Director	1970-71
	Membership Director	1972
Garland Weeks	Market Analyst	1970-71
David Hales	Market Analyst	1971-72
Cathy DeBord	Market Secretary	1971-73
Shirley Baker	Secretary/Office Manager	1972-80
Gery Donnell	Market Analyst	1972-73
Charles E. Ball	Executive Vice President	1972-88
Jim Peterson	Executive Assistant	1973
	Membership Director	1975-84
Peggy Sherron	Receptionist	1973-
Vi Jean Fuglaar	Market Secretary	1973-74
Mike Riethmayer	Membership Director	1973
Brenda Higley	Administrative Assistant	1974-
Gary Culp	Membership Director	1974-77
Richard McDonald	Executive Director	1974-88
	Executive Vice President	1988-
Steve Munday	Information Director	1974
Joe Wood	Market Analyst	1974-78
Linda Seitz	Secretary	1974-77
Jim Gill	Market Director	1974-
Glen Starcher	Information Director	1975-76
Sherri Evans	Clerk/Typist	1976-77
Lisa Neal	Secretary	1976-77
John McCarroll	Information Director	1976-77
Steve Cornett	Information Director	1977-78
Pam Cosby	Secretary	1978-80
Barbara Steinle	Secretary	1978-80
John Kuker	Market Analyst	1978-79
Anne Ilse Anderson	Director of Nutrition	1978-83
Joyce Calhoun	Secretary	1978-84
Robert Price	Market Analyst	1978-80
Lynn Heinze	Communications Director	1979-80
Chris Hyndman	Market Analyst	1979-81
Charlie Christopher	Membership Representative	1979-83
Lee Wilcox	Communications Director	1980-84
Sharon Pinson	Accountant	1980-88
Sharon Hartman	Membership Secretary	1980-82
Thomas Horndt	Market Analyst	1980-83
Patti Jones	Communications Secretary	1981-83
Fran Blessen	Computer Operator	1981-83
J. B. Douglas	Market Analyst	1981-82
Gayle Huff	Accounting Clerk	1982-84
Charis Young	Membership Secretary	1982-84
Larry Fullen	Market Analyst	1982-84
Amy Swedberg	Insurance Secretary	1983-85
Stan Price	Market Analyst	1983-86
	Market Analyst	1988-89
Sheila Land	Communications Secretary	1983-84
Sheryl Brown	Govt. Affairs Secretary	1984-85
Ron Banks	Membership Director	1984-87
Steve Johnson	Market Analyst	1984-89
Madell Garrett	Accounting Clerk	1984-89
Kay Sheffield	Membership Secretary	1984-
Cheryl Smith	Communications Secretary	1984-86
Becky Hefner	Insurance Secretary	1984-89
Robert Hughes	Market Analyst	1985
Burt Rutherford	Communications Director	1985-
Ross Wilson	Govt. Affairs Director	1985-
Patti Adams	Market Assistant	1985-86
Shannon Standley	Govt. Affairs Secretary	1985-
Mike Boswell	Market Analyst	1985-89
Penny Hammack	Communications Secretary	1986-87
Angela Knapp	Market Secretary	1986-
Bill Carrothers	Market Information Systems	1987
Derrell Havins	Safety Manager	1987-
Lisa Andrew	Communications Secretary	1987-88
Pat Renner	Part-time Secretary	1987-88
Don Clift	Membership Manager	1987-
Ilene Tuey	Communications Secretary	1988-92
Matt Forrester	Accountant/Office Manager	1989-90
Brent Koller	Market Analyst	1989-
Tom McDonald	M-Net Manager	1989-
Jerry McCarty	Market Analyst	1989-
Carol Nordquist	Accounting Secretary	1989-
Glenna Doose	Insurance Secretary	1990-91
Nathan Bailey	Accountant/Office Manager	1990-91
Brenda Reeder	Administrative Secretary	1991-92
Judy Heringa Luna	Insurance Secretary	1991-
Connie Tommerup	Controller	1992-
Carla Walker	Administrative Secretary	1992-

NOTES AND BIBLIOGRAPHY

Chapter 1, "Feedyards and Forefathers," is condensed from a chapter, by Charles E. Ball and Steve Cornett, in *Cattle Feeding: A Guide to Management*. Amarillo: Trafton Printing, 1990.

Much of Chapter 6, "Entrepreneurs and the Boom," and Chapter 7, "The Wreck," is excerpted from *Texas Cattle Feeding: The Genesis of An Agricultural Giant.* Amarillo: Texas Cattle Feeders Association, 1977.

Chapter 8, "A New Association," Chapter 9, "Major Issues," Chapter 10, "TCFA Leaders," and Chapter 11, "The Professionals," are based on information in the files of the Texas Cattle Feeders Association, Amarillo, and on the author's knowledge and experiences.

Personal Interviews

In 1977, while researching for *Texas Cattle Feeding: The Genesis of An Agricultural Giant*, published by Texas Cattle Feeders Association in 1977, Steve Cornett, Communications Director of TCFA, conducted a number of interviews which provided information for this book. They included the following:

Fehleison, Mary Alyce. Sisterdale, Texas, 1977.
Moser, Norman. DeKalb, Texas, 1977.
Riggs, John K. College Station, Texas, 1977.
Simmons, Tom B. Jr., Lubbock, 1977.
Snyder, R. L. Clayton, New Mexico, 1977.
Weeman, Roy. Fort Worth, 1977.

During 1991-92, the author conducted a number of interviews, including the following:

Armstrong, John B. Kingsville, Texas, 1992.
Barnes, George S. Hobart, Oklahoma, 1992.
Bennett, Paul D. Littlefield, Texas, 1992.
Bergsma, Lloyd. Amarillo, 1991.
Black, A.L. Friona, Texas, 1991-92.
Bogle, Bill. Dexter, New Mexico, 1991.
Boog-Scott, J.E. Jr. Cleburne, Texas, 1992.
Bralley, Arthur P. Amarillo, 1991.
Cummings, Ernest. San Marcos, Texas, 1991.
Davidson, Jim. Santa Fe, New Mexico, 1991.
Eoff, Luther. Oklahoma City, 1991.
Foley, Marshall R. Eufaula, Oklahoma, 1991.
Freeman, Jack M. Texhoma, Oklahoma, 1992.
Johnson, M.T. Amarillo, 1991.
Keenan, G. Wendell. Russellville, Arkansas, 1992.
McAndrews, James. Iowa City, Iowa, 1991-92.
McCasland, Z A. Hereford, Texas, 1992.
Merrill, Lee. Clovis, New Mexico, 1992.
Mock, Paul. Altus, Oklahoma, 1991.
Ridley, Fred, Sr. Oklahoma City, 1991.
Riggs, John K. College Station, Texas, 1991.
Sanders, C.T. "Tad." Kansas City, Missouri, 1991.
Sartwelle, J.D. Houston, 1992.
Schoeb, Jerry. Cherokee, Oklahoma, 1991.
Simmons, Tom B. Jr. Lubbock, 1992.
Simpson, Robert L. Jr. Eufaula, Oklahoma, 1991.
Smith, Earnest. Roswell, New Mexico, 1992.
Thomas, J. Harper. Chickasha, Oklahoma, 1992.
Tollett, Marvin. Sudan, Texas, 1991.
Worley, Victor. Portales, New Mexico, 1992.
Wright, Caddo. Lubbock, 1991.

Bibliography

The references noted below normally would appear in footnotes on the same pages with the facts reported. In this book, however, it was decided that such scholarly format would slow or impede the reading. All historical facts, in addition to those experienced or observed personally by the author, were scrupulously obtained from the interviews listed on this page and from the publications listed below.

Allen, Louis F. *American Cattle: Their History, Breeding and Management.* New York: Orange Judd, 1868.

Ammerman, C. B. and Henry, P. R. "Ruminant Nutrition: A Century of Progress." Ithaca: *The Cornell Veterinarian*, January 1985.

Armsby, Henry Prentiss. *A Manual of Cattle Feeding.* New York: John Wiley and Sons, 1880.

Ball, Charles E. "Atomic War on Insects." *The Saturday Evening Post,* September 9, 1961.

Ball, Charles E. and Cornett, Steve. "Cattle Feeding in the United States." *Cattle Feeding: A Guide to Management.* Amarillo: Trafton, 1990.

"Beef Cattle Investigations in Texas, 1888-1950," Bulletin 724. College Station: Texas Agricultural Experiment Station, 1950.

Berglund, Roger, et al. *Forty Years of Improvement in Cattle Feeding.* Minneapolis: Miller Publishing Co., 1969.

Bidwell, Percy Wells and Falconer, John L. *History of Agriculture in the Northern United States*, 1620-1860. Washington: Carnegie Institution, 1925.

Brayer, Garnet M. and Herbert O. *American Cattle Trails*, 1540-1900. Bayside: American Pioneer Trails Association, 1952.

Browne, Charles A. et al. *Liebig and After Liebig*. Washington: American Association for the Advancement of Science, 1940.

Burk, L. B. "History of Meat Grading," a paper for Agricultural Marketing Administration, U.S. Department of Agriculture, Washington, D.C., 1944.

Cattle Raisers Association of Texas, Minutes, 1895-1905. Library, Texas and Southwestern Cattle Raisers Association, Fort Worth.

Cattleman, The. Fort Worth: Cattle Raisers Association of Texas, 1914-1970.

CME News Release, "25 Years of Innovation at CME." Chicago: Chicago Mercantile Exchange, 1989.

Cotton, J. S. and Ward, W. F. "Economical Cattle Feeding in the Corn Belt." USDA Farmers Bulletin 588. Washington, June 1914.

Cox, James. *Historical and Biographical Record of the Cattle Industry and the Cattlemen of Texas*. Saint Louis: Woodward & Tiernan, 1895.

Cross, H. R. and Harris, J. J. "History of Meat Grading," a paper for Department of Animal Science, Texas A&M University, 1990.

El Paso (Texas) Times. El Paso: Several articles, April 5, 1953; September 10, 1967; June 27, 1971; September 27, 1989.

Farm Journal. Philadelphia: Farm Journal, Inc., 1954-1972.

Fowler, Stewart H. *The Marketing of Livestock and Meat.* Danville: The Interstate, 1961.

Grain Sorghum Research in Texas...1970. College Station: Texas Agricultural Experiment Station, 1971.

Gray, Lewis Cecil and Thompson, Esther Katherine. *History of Agriculture in the Southern United States to 1860*. Washington: The Carnegie Institution, 1933.

Green, Donald E. *Panhandle Pioneer*. Norman: University of Oklahoma Press, 1979.

Green, Donald E. *A History of Texas Tech University's College of Agricultural Sciences*, 1925-1975. Lubbock: Texas Tech Press, 1977.

Green, Donald E. *A History of the Oklahoma State University Division of Agriculture.* Stillwater: Oklahoma State University, 1990.

Haney, Robert L. *Milestones Marking Ten Decades of Research.* College Station: Texas Agricultural Experiment Station, 1989.

Harris, Joseph. "On Stall Feeding Cattle and Sheep." Report of the Commissioner of Agriculture. Washington: The Government Printing Office, 1863.

Henlein, Paul C. *Cattle Kingdom in the Ohio Valley*, 1783-1860. Lexington: The University of Kentucky Press, 1959.

Henry, W. A. *Feeds and Feeding*. Madison: W. A. Henry, 1898.

Hopkins, John A. *Economic History of the Production of Beef Cattle in Iowa*. Iowa City: The State Historical Society of Iowa, 1928.

Kay, Steve. "The Beef Industry's Top Ten," *Directions*. Denver: National Cattlemen's Association, 1988.

Kay, Steve. "The Top 25," *Directions*. Denver: National Cattlemen's Association, 1991.

Kunkel, H. O. *Nutritional Science at Texas A&M University, 1888- 1984*. College Station: Texas Agricultural Experiment Station, 1985.

Lacy, Mary G. "Food Control During Forty-Six Centuries," an address before the Agricultural History Society, Washington, D.C., March 16, 1922.

Lambert, Roger. "Drought Relief for Cattlemen; The Emergency Purchase Program of 1934-35." Panhandle-Plains Historical Review. 45 (1972): 21-35.

M'Combie, William. *Cattle and Cattle Breeders*, 1875. Parsons: McClain Publishing Co., Reprinted 1983.

Magner, D. *Magner's Farm and Stock Book and Complete Instructor*. D. Magner, 1902.

Marion, Paul T. "Feed Additives in Cattle Rations." Speech to First Annual Cattle Feeding Conference, Lubbock, sponsored by Cattle Feeders Division, TSCRA, December 6, 1961.

May, Irvin, Jr. "Welfare and Ranchers: The Emergency Cattle Purchase Program and Emergency Work Relief Program in Texas, 1934-1935." West Texas Historical Association Year Book. 47 (1971): 3-19.

May, Michael L. "Beef Grading – Past and Present," a paper for Agricultural Marketing Service, U.S. Department of Agriculture, Washington, D.C., 1977.

McCoy, John H. *Livestock and Meat Marketing*. Westport: AVI Publishing Company, 1979.

McCoy, Joseph G. *Historic Sketches of the Cattle Trade of the West and Southwest.* Kansas City: Reton, 1874. Reprint, Washington: The Rare Book Shop, 1932.

McDonald, Richard. "Regulatory Agencies." *Cattle Feeders Annual*. Amarillo: Texas Cattle Feeders Association, 1976.

Meat and Poultry Inspection. Washington: National Academy Press, 1985.

Morrison, F. B. *Feeds and Feeding*. Ithaca: The Henry-Morrison Company, 1928.

Nall, Garry L. "The Cattle Feeding Industry on the Texas High Plains." *Southwestern Agriculture: Pre-Columbian to Modern*. Edited by Henry C. Dethloff and Irvin M. May, Jr. College Station: Texas A&M University Press, 1982.

Nimmo, Joseph Jr. "*The Range and Ranch Cattle Business of the United States*." Report on the Internal Commerce of the United States. U.S. Treasury Department, May 6, 1885.

Nourse, Edwin G.; Davis, Joseph S.; Black, John D. *Three Years of the Agricultural Adjustment Administration*. New York: Da Capo Press, 1971.

Parker, Evelyn K. "Commodity Exchange Act Amendment of 1958: Legislative History." Washington: Commodity Futures Trading Commission, 1981.

Pate, J'Nell L. *Livestock Legacy, The Fort Worth Stockyards*, 1887-1987. College Station: Texas A&M University Press, 1988.

"Principles of Stock Feeding and Some New Mexico Feeding Stuffs," Bulletin 17. Las Cruces: New Mexico Agricultural Experiment Station, 1895.

Prose and Poetry of the Live Stock Industry of the United States. Denver and Kansas City: National Live Stock Historical Association, 1904.

Rassmussen, Wayne D. *Agriculture in the United States*, A Documentary History. New York: Random House.

Ross, John F. and Leidigh, A. H. "Cereal Experiments in the Texas Panhandle," BPI Bulletin 283. U.S. Department of Agriculture, 1913.

Russell, Howard S. *A Long, Deep Furrow–Three Centuries of Farming in New England*. Hanover: University Press of New England, 1976.

Schlebecker, John T. *Whereby We Thrive, a History of American Farming, 1607-1972*. Ames: The Iowa State University Press, 1975.

Schlebecker, John T. *Cattle Raising on the Plains*, 1900-1961. Lincoln: University of Nebraska Press, 1963.

Skaggs, Jimmy M. *Prime Cut: Livestock Raising and Meatpacking in the United States*, 1607-1983. College Station: Texas A&M University Press, 1986.

Sinise, Jerry. "The History of Cattle Feeding in Texas." Unpublished manuscript for TCFA, 1975.

Stewart, E. W. "Cattle Feeding in New York." Report of the Commissioner of Agriculture. Washington: The Government Printing Office, 1877.

Taylor, Joe F. "Federal Aid to the Farmers in the Panhandle of Texas." M. A. Thesis in History. Canyon: West Texas State College, 1940.

Texas Cattle Feeding: The Genesis of An Agricultural Giant. Amarillo: Texas Cattle Feeders Association, 1977.

Thompson, James Westfall. *A History of Livestock Raising in the United States*, 1607-1860. USDA Agricultural History Series 5. Washington, 1942.

Vansickle, Joe. "Why 'New' Is 15 Years Old." Minneapolis: *Beef*, March 1992.

Ward, Clement E. "Packer Control of Inventory – What It Means to You." *Cattle Feeders Annual*. Amarillo: Texas Cattle Feeders Association, 1990.

Wheeler, David. "The Beef Cattle Industry of the United States: Colonial Origins," *Panhandle-Plains Historical Review*. 46 (1973): 54-67.

Whitaker, James W. *Feedlot Empire: Beef Cattle Feeding in Illinois and Iowa, 1840-1900*. Ames: Iowa State University Press, A Replica Edition, 1975.

Wilbur, John L., DVM. "Historical Data: Texas Animal Health Commission." Paper prepared for TAHC, Austin, May 1978.

World Book Encyclopedia. Chicago: Field Enterprises Educational Corporation, 1961.

Yearbook of Agriculture, 1935. Washington: U.S. Department of Agriculture, 1935.

Zimmerman, William David. "Live Cattle Export Trade Between United States and Great Britain, 1868-1885." *Agricultural History*. 36 (January 1962): 46-52.

INDEX

A

B

C

D

E

F

G

H

I

J

K

L

M

N

O

P

Q

R

S

T

Illustration for 1978 Cattle Feeders Annual, designed for TCFA by cartoonist Stan Lynde.